Delivering Cons
Operations Building
exchange (COBie) in
GRAPHISOFT ARCHICAD

E. William East and Robert Jackson

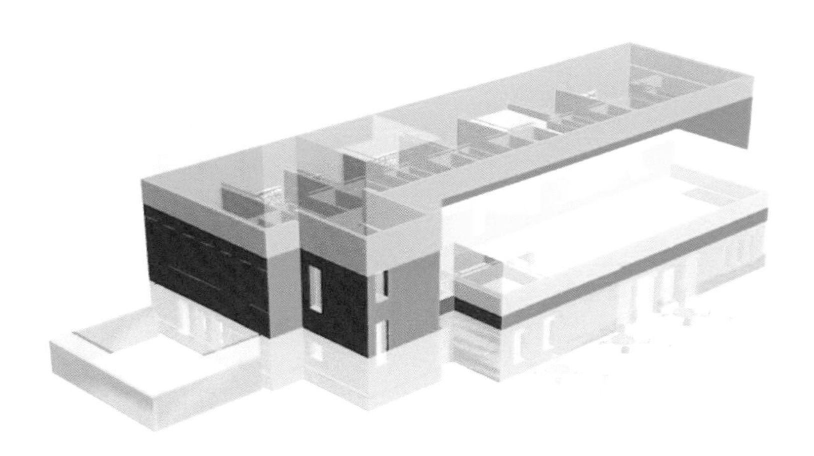

COPYRIGHT NOTICE

Printed in the United States of America

ISBN 978-1-365-26840-3

For permission requests, write to the publisher, addressed "Attention: Permissions Coordinator," at the address below.
Prairie Sky Consulting LLC
1104 Parkview Drive
Mahomet, IL 61853

www.prairieskyconsulting.com

ACKNOWLEDGEMENTS

The authors would like to thank GRAPHISOFT for their technical support, detailed review, and sponsorship of the production of this book. Without their support for transparent and open COBie testing, this project would not have been possible.

TERMS OF USE

Your use of this book constitutes your explicit agreement to the terms of use listed below.

All content in this book, and all referenced content, is provided for informational purposes only. The authors make no representations as to the accuracy or completeness of any information in this book or referenced materials. Prairie Sky Consulting, Bond Bryan Architects, Bond Bryan Digital, and the authors of this book shall not be liable for any errors or omissions in this information nor for the availability of referenced information. These parties will not be liable for any losses, injuries, or damages from the display or use of this information.

ABOUT THE AUTHORS

E. William "Bill" East is a serial innovator, whose latest invention, COBie, went from concept to world-wide standard in less than a decade. In his capacity as COBie Academy Director, Bill provides on-line university courses where professionals learn the COBie standard and anticipate its impact on their practice. As the Owner of Prairie Sky Consulting®, Bill helps contractors and owners eliminate waste through better information management. Bill recommends COBieScoreCard℠ to help manage COBie adoption.

Robert "Rob" Jackson is a world-known ARCHICAD expert and COBie enthusiast. Rob has written many blogs about best-practices for obtaining high-quality IFC and COBie data from ARCHICAD. In this book, Rob completes the architectural design of a brand new project and documented the steps needed to deliver COBie data in ARCHICAD. As Associate Director at Bond Bryan Digital, Rob is able to assist others in understanding and delivering open workflows on real life projects.

See Bill's Curriculum Vitae at www.linkedin.com/in/williameast.
Contact Bill at bill.east@prairieskyconsulting.com.

Find out more about Bond Bryan Digital at www.bondbryan.com/digital.
Contact Rob at bim@bondbryan.co.uk.

TABLE OF CONTENTS

TABLE OF CONTENTS (cont.)

TABLE OF FIGURES

TABLE OF FIGURES (cont.)

PREFACE

This book explains how architects can use GRAPHISOFT's ARCHICAD 19 (Build 4013) to obtain high quality COBie output. The book includes tips and tricks for creating COBie efficiently gleaned from this effort and other real-life examples. The goal of highlighting these issues is to ensure that users can clearly understand what is easy, and not as easy to do, using the software demonstrated in this book. These issues are in the process of being reviewed by GRAPHISOFT. In some places areas that may be improved, or simplified, have been identified in the book's Appendix. Users should review the release notes for subsequent build numbers to determine whether any of the issues identified have been addressed since the publication of this book.

Before beginning, GRAPHISOFT's guide should be the first document for any ARCHICAD user who needs to understand and deliver COBie [GRAPHISOFT 2015]. There are also guides for ARCHICAD from version 16 onwards available from GRAPHISOFT. The address for to this information may be found at the end of the Reference chapter.

In addition to the GRAPHISOFT Guide, references in this book provide vital information for those looking to expand their knowledge of COBie. The two most important references are the COBie standard itself and the files containing the example building used in this book. The US National Building Information Model Standard (NBIMS-US V3) that defines COBie is available here [NIBS 2015]. The link to the East Dormitory project files used to test the ideas developed in this book are available here [East 2016]. Readers should download this information for ready reference.

Readers interested in a deeper understanding of the application of COBie to design and construction may want to join the University of Florida's COBie Academy [University 2016].

Ultimately, the authors' goal in writing this book is to share the experiences of software users and their partners, stakeholders, and clients who use the COBie data they produce.

THE COBIE STANDARD

The first COBie specification was published in 2007 [East 2007]. The purpose of COBie is to provide an alternative to paper-based construction handover documents found in virtually every building in the world. Figure 1 illustrates the extent of such documents for one single building.

Figure 1 Paper Handover Documents

COBie was not meant to solve every information exchange problem facing our industry today. COBie was developed as a very specific tool to solve a very specific problem. COBie was designed to capture building information about building components that have moving parts; components that require scheduled maintenance. COBie can be described, simply, as "information about Pump-5 in Room-3".

Although the requirement for COBie was first identified as part of a United States government project, the COBie standard is not a government-driven standard. COBie development has been accomplished as a collaboration between technical teams in several countries, software companies, building owners, and facility managers.

Throughout the development of COBie, a collaborative approach was used because the COBie team understood that the only way for owners to obtain accurate handover information was to engage the software systems that created, updated, and delivered that information every day, on every project. During the first decade of work, over thirty different software programs participated in the discussion and implementation of COBie [East 2014]. The diversity of these programs from planning and programming, to architected design, to engineering design, to construction management, to facility management, runs the entire course of the building life-cycle.

COBie is not meant to be completed by a single party in a building's life-cycle. Planners, designers, engineers, contractors, and commissioning agents each have their unique part to play. Historically, these professionals did their jobs on paper. With COBie, it is now the job of their software systems to help these professionals do that same job in a consistent, streamlined process. Having the COBie common data structure assists each party to do their job more easily. Users no longer have to begin by recreating, transcribing, copying, and or entering that same information each time it is needed.

A schematic diagram of the organization of COBie data is shown in Figure 2 [East 2009]. There are three major groups of information in COBie, organized by color coding. The first set, shown in blue, is information provided by designers. Designer's information is comprised of two major categories: spaces and components. The spatial structure identifies the names of each space within a building and the organization of those spaces. The physical organization of spaces is based on the "facility" (or building) and the "floor" (or story). Groups of "spaces" (or rooms) are organized into "zones" (and sub-zones).

The second category of design information identifies equipment assets to be maintained and operated. These assets are individually listed as "components" (or products and equipment). Given that most products are not engineered on-site, but are off-the-shelf manufactured products, components are identified by "type". Groups of "components" (or products and equipment) are grouped into "systems" (and sub-systems").

As noted before, designers are not responsible for the complete set of COBie data. The other major contributors to COBie are contractors and subcontractors working during the "build" phase of the project. This could also include work accomplished during the commissioning phase of the project. The colored orange shape in Figure 2 illustrates the information provided during the build phase.

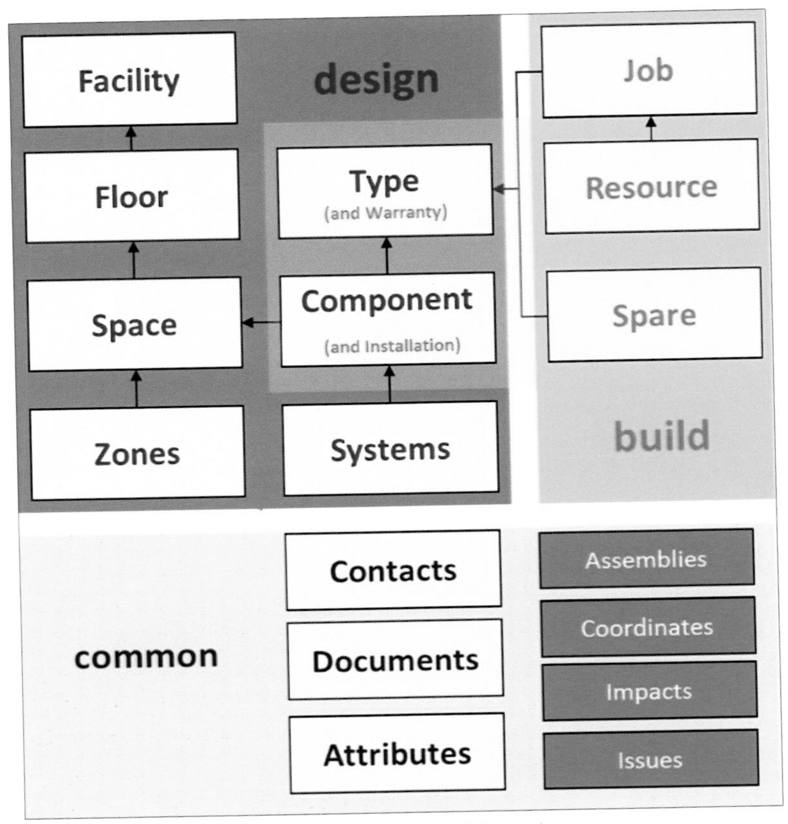

Figure 2 COBie Data Schematic

Notice the overlapping colored blocks on the "type" and "component" parts of the COBie data schematic. For such information, both designers and builders have a COBie contribution to make. The designer is responsible to provide the outline of the types of equipment to be provided and the spatial location of each of these products and pieces of equipment in the building. The builders (contractor, subcontractor, and commissioning agent) are responsible to fill-in the outline of products and equipment installed in the building. The builder also identifies the Operations and Maintenance (O&M) jobs required to keep the building operating efficiently. The resources needed do perform those jobs and on-site spares (lubricants and replacement parts) required are also identified.

The major requirement for common data is that for "contacts," "documents," and "attributes." Contacts provide the information for companies such as designers, engineers, contractors, subcontractors, and product manufacturers. Documents reference information provided in files that accompany a COBie deliverable. During design, such documents include design reports and contract documents. During construction, COBie documents are primarily produced during the construction submittal and approval process. In Figure 2, "common" information is information that may be provided throughout the project as needed.

COBie requires only the most minimal set of information needed for building handover across any category of buildings, spaces, components. Attributes are a very important part of the COBie common data sets because they hold detailed information beyond that minimal set.

Since the purpose of COBie is to provide facility operators and maintainers the ability to do their jobs, the exact location and dimensions of the building's components are not needed. Traditional Operations and Maintenance (O&M) practice only requires the identification of room (or "space") from which the component (or "equipment") is operated. As a result, COBie data does not require geometric information. While geometric information is critical to the production of drawings and the design and construction coordination process, COBie's ability to capture boxes, lines, and points is inadequate for even the simplest of rectangular buildings. In the current COBie standard, such simple constructs may be provided but are of little value. These "coordinates" appear in the green "common" block of Figure 2.

The COBie standard is a product of the buildingSMART alliance through its publication of the United States National Building Information Model Standard (NBIMS-US). The current version, version 3 [NIBS 2015], is freely available following free registration. The COBie standard is found in Chapter 4.2 of NBIMS-US V3.

Everyone working with COBie should download a copy of NIBMS-US V3 for reference.

This book describes the requirements for the delivery of COBie data during the early stage of architectural design only. Later books in this series will describe how COBie data is created during coordinated design, construction, and commissioning using specific software that is helpful to each of those project phases.

CODE, COMMENTARY, AND SPECIFICATION

At its core, design and engineering practice requires three types of documents. These documents define and enforce a specific minimum standard of care. It is our shared understanding of that standard of care that allows our complex industry to function. Once all parties to a contract understand what is to be provided, contracts can be issued with minimal litigation and maximum efficiency.

The first of these types of documents are Codes and Standards. These documents contain technically-oriented language meant to provide the most precise and unambiguous statement of what is to be done.

The second of these are Commentaries and Standards of Practice documents. Commentaries describe the application of a referenced Code or Standard to different conditions. Commentaries are written in professionally-oriented language that directs the reader to understand the applications and limitations of the referenced Code or Standard.

In contrast to Commentaries, Standards of Practice documents are often not published by the original Code or Standards body. These documents may also be adapted from general Commentaries to describe specific regional, national, or local project contexts.

The original COBie Guide was developed by the first author in 2012 [East 2012a]. Today many federal, state, and local governments, as well as many private owners, have adapted their own versions of that original COBie Guide.

The third of these documents are Contract Specifications. Specifications define what is to be accomplished. Specifications are written in legally-oriented language. Specifications give the details of specific deliverables that are to be provided against the cited Code and Commentary or Standard of Practice document.

The original COBie Specification was developed by the first author in 2012 [East 2012b]. As with the COBie Guide, many owners have adapted that example to reflect their own requirements.

It is critical to note that COBie should be developed as a stand-alone specification and not combined with "Building Information Modeling" contract requirements. While technology may assist in meeting COBie requirements, as explained in this book; it is not the specifier's job to direct designers, engineers, and contractors to select specific technology and/or proprietary software when there are potentially many products that may correctly deliver COBie data.

CUSTOMIZING COBIE

The COBie standard is published by the buildingSMART alliance, a council of the US National Institute of Building Sciences (NIBS). While the COBie standard maybe freely used for purposes both commercial and non-commercial, there are restrictions to its use. These restrictions are based on the Creative Commons License under which COBie was first developed [NIBS 2012].

According to the COBie standard [NIBS 2014], there are three ways that COBie may be customized for regional, local, or project-based usage. The first is that the classification system used for specific COBie data should be specified to support local and/or project requirements. While COBie in the US defaults to the Construction Specification Institute's OmniClass category codes, COBie in the UK uses Uniclass as a default.

In this book, the dormitory project was developed primarily for US COBie usage, and therefore contains OmniClass Classification references. The UK equivalent file would simply substitute Uniclass for OmniClass.

In addition to the selection of alternative classification schemes, COBie may be customized by further limiting the set of building assets considered to be "managed assets." NBIMS-US V3 specification provides a complete set of the elements in an overall building information model that are to be excluded from a standards-compliant COBie deliverable. For example, COBie does not consider structural elements (walls, floors, columns) or fluid transport elements (ducts, pipes, and wires) to be "managed assets." While there is certainly a requirement for the collection and exchange of such information on a building project, that information is explicitly outside the scope of COBie as defined in NBIMS-US V3.

The final customization allowed by COBie is the identification of the attributes, or properties, to be provided for the required set of "managed assets." In this regard, US COBie usage follows industry practice regarding the delivery of such information. When information is normally available as part of a typical design and construction document (such as a drawing schedule or installed equipment list), that information should be provided in COBie. Other countries and regions may take more prescriptive approaches that may, or may not, follow industry standard practice. Such approaches include the establishment of arguably arbitrary standard product libraries, Levels of Detail, or Levels of Development.

While the example project described this book, was developed primarily for COBie-US usage, if there are specifically required COBie-

UK attributes these have been added. Such fields are identified and referenced where appropriate in the text.

The reader should note that these customizations do not require changing or reordering of any COBie spreadsheet Tabs and Columns allowed. To change or reorder any COBie spreadsheet Tabs and/or Columns is not explicitly disallowed by the COBie standard.

To reinforce the fact that these three types of customizations described above are the only permitted types, COBie is provided under Creative Commons Licensing [NBIMS 2012]. Under the terms of the "Attribution-NoDerivs 3.0 Unported" license anyone may share and use the COBie standard in any format or media. COBie may be used for any purpose, including commercial use. There are also several restrictions on the use of COBie. The first of these is that any use of COBie must reference the COBie standard provided by the buildingSMART alliance.

This restriction ensures that there are not multiple, varying specifications for COBie throughout the world. Having a single standard allows all parties using it to simplify their operations since they no longer need individual programming to support proprietary data exchange formats. An example of a possible violation of this standard is the publication of previous versions of COBie specifications on websites of organizations other than that of the buildingSMART alliance.

The second restriction that if COBie is used differently than specified, then it must not be called COBie. COBie-UK, for example, requires the use of COBie "for all" types of projects, including non-building projects. The use of COBie for projects other than buildings, is outside the scope of the COBie standard, requiring the identification of COBie-UK. COBie-UK's emphasis on capturing life-cycle costs and carbon footprint are also extensions of the original COBie scope and are not addressed in NBIMS-US V3.

As this book pertains only to the application of COBie to a single building, the differences between COBie and COBie-UK will be identified as they occur when using the software itself.

Agreements among some software companies may also have gone beyond the COBie standard. Such changes, called "implementers' agreements," often add new columns to the right of standard COBie columns. As with any deviations from the standard, the resulting files must be called something other than "COBie." Discussion of implementers' agreements are not included in this book.

COBIE PRESENTATION FORMATS

A data schema defines the required content of a set of information, but it does not define the way that such information is delivered.

A schematic of the COBie data structure was presented in Figure 2. The technical specification of COBie is formally called the COBie Model View Definition (MVD). The COBie MVD can be found in Chapter 4.2 of the United States National Building Information Model Standard (NBIMS-US) V3. The unofficial, web-based version of the COBie MVD has also been published by the buildingSMART alliance for easy reference based on the most recent ifcDocs format [Chipman 2013].

It should be noted that information that of the time this book's publication found on the buildingSMART international website entitled "Basic FM Handover MVD" has not been valid since 2012. While repeated requests to remove this depreciated information have been made to members of the buildingSMART international, the reason for their publication of outdated information is unknown.

NBIMS-US V3 defines the physical formats in which COBie data can be delivered. There are four formats allowed. The first is the familiar COBie spreadsheet. The mapping between the formal COBie schema (in MVD format) and the COBie spreadsheet is provided in NBIMS-US V3. The second format is STEP Physical File Format (SPFF). This file is often incorrectly referred to as an "IFC file". The third format is an ISO-compliant representation of an SPFF called ifcXML. The fourth and final format provides sub-schema that allow mini-exchanges of parts of an overall COBie data set. This final format is called COBieLite.

Two COBie presentation formats are discussed in this book, SPFF and spreadsheet format. Files in this format will referred to as "COBie SPFF-files" when referred to in this book. Files in spreadsheet format will be referred to as "COBie spreadsheets" or "COBie Files." To consistently describe the building information contained in COBie spreadsheets, this book employs the following conventions:

- "COBie spreadsheet" – technically the file containing multiple spreadsheets is called a "workbook;" however, this technically correct description is not typically used in practice. In this book, the term "COBie spreadsheet" refers to a single file, typically with a file extension of ".xlsx," comprised of multiple spreadsheet "Tabs."

- "COBie Tab" – an individual spreadsheet within a COBie workbook. COBie Tab's will often be referred to by their names, as in "COBie.Space" or "COBie.Component."
- "COBie Table" – there is a close connection between the structure of the COBie spreadsheet (i.e. workbook) and a relational database. As a result, many refer to a given COBie Tab as a COBie Table.
- "COBie Column" – an individual column of data within a given COBie Tab. With the exception of the COBie.Instruction Tab, the first row in every COBie Tab provides the COBie Column name. COBie Columns will often be referred to by their names, as in "COBie.Space.Name." The position of the COBie Column may also be identified; however, the typical usage will be to refer to the name of the column.
- "COBie Field" – there is a close connection between the structure of the COBie spreadsheet (i.e. workbook) and a relational database. As a result, many refer to a given COBie Column Name as a COBie Table Field, or simply COBie Field.
- "COBie deliverable" – a package of files provided to meet a specific contract specification. At the architectural design stage of a project, this package will typically include: (a) the COBie spreadsheet (workbook) file, (b) the set of files referenced in COBie.Document, and (c) a COBie quality control report showing zero errors.

As noted in an earlier chapter, COBie is a data model that can support life-cycle information exchange about the rooms and equipment in a building. Not all COBie data is meant to be delivered in any specific COBie deliverable. An essential convention in COBie is, therefore, the idea of "NULL" data. The identification of NULL data means that a given COBie Field does not have a value.

COBie requires a positive assertion that data is, or is not, provided for every column found in a given COBie Tab. This ensures that the receiver of the data will understand if data is missing, or if that data was not provided (i.e. NULL). To help simplify the conversion of COBie data in multiple formats (.ifc, .xlsx, .xml) the designation of NULL in each format is required.

In the STEP presentation of COBie data NULL is identified by the "$" character for text and number fields. The value of January 1, 1900 is used in a STEP file to deliver NULL date information. Since such constructs will be misinterpreted by spreadsheet programs a different value is required to indicate NULL. In COBie spreadsheets, the value for NULL is "n/a".

For those not interest in the full explanation of the NULL value, "n/a" can be described as "not applicable" or "not available"; however, the technical meaning of this value is NULL. Providing actual values indicating that information will be provided later, such as "TBD" for "To Be Determined," are not NULL values and such will be incorrectly interpreted as actual values by receiving software. A COBie best-practice is to always use "n/a" in a COBie spreadsheet if data is not applicable or not available for a given COBie spreadsheet.

TESTING PROCESS

In less than a decade, the Construction-Operations Building information exchange gained acceptance and has become the basis for national (and defacto) standards throughout the world. One of the reasons for the success of this effort has been the care and attention paid to open and public software testing.

COBie software testing has evolved through several stages mirroring the increasing ability of software to produce COBie data. At the start, software companies' effort and testing was focused simply on producing or consuming data in one of the specified formats. Once import and export capabilities were publically demonstrated, testing focused on providing the correct data in the correct format. Now that this milestone has been reached (for some of vendors claiming COBie compliance), the aim of COBie testing is to assist designers and builders to most efficiently use that software to produce high-quality COBie deliverables.

A testing process that can be replicated outside a once-every-few-years conference room or laboratory setting requires three key features. These features are: (1) the use of a control data set or "test building," (2) the verification that the information provided meets the technical requirements of the standard, and (3) the validation that the information provided is correct.

Test Building

Rather than a use a set of design information provided by a given vendor, it is critical that the test building be independently developed. The building's information must be available for full public inspection. Most importantly, the files must be made freely available so that anyone who wishes to do so is able to replicate the results of the test.

The control model used in this book was developed from a governmental agency's standard design drawings for a small, two-story dormitory. These drawings have been available free-of-charge through the buildingSMART alliance since 2013 [East 2013]. Figure 3 illustrates the ground floor plan for this building.

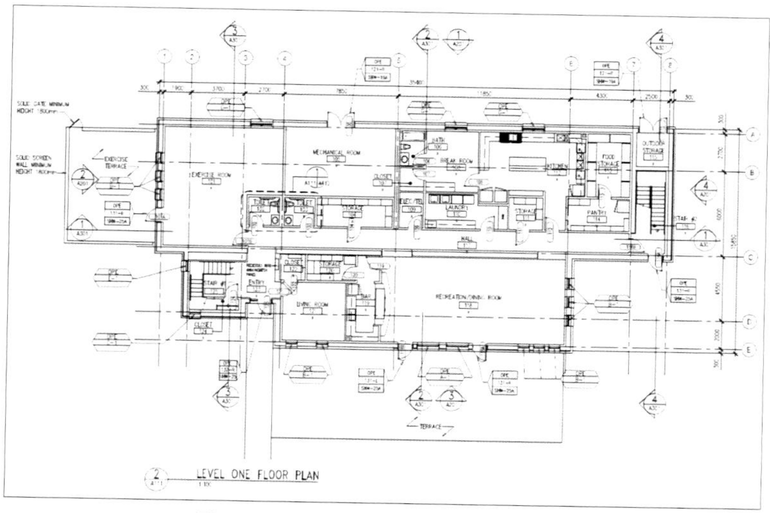

Figure 3 Dormitory Building Floor Plan

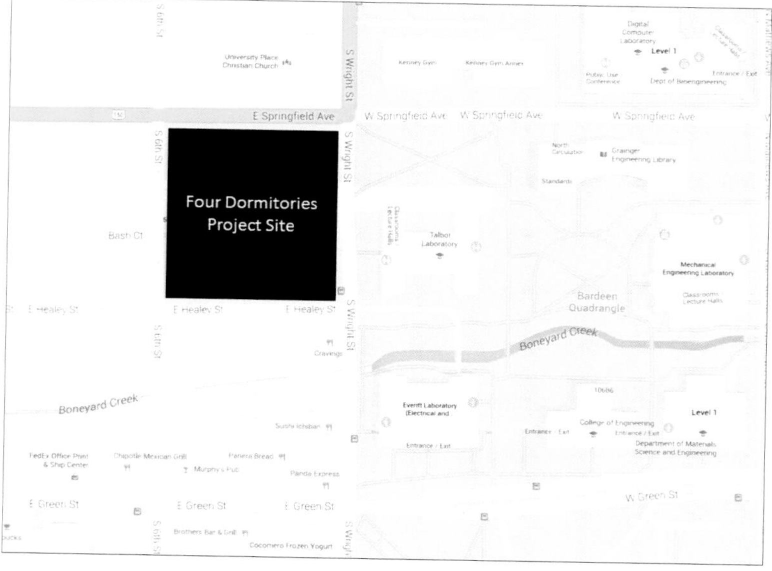

Figure 4 Four Dormitories Project Site

For the purposes of COBie testing, the example dormitory model has been re-purposed as a fictional example of one of four dormitories that is planned to be built at the campus of the University of Illinois. The specific, but fictitious, location chosen for the "Four Dormitory Project" is in a parcel adjacent on the North West corner of the intersection of East Healy Street and South Wright Street in Champaign, IL, USA. Figure 4 provides an illustration of the location of the project site. To the east of the site is the engineering campus of the University of Illinois. Four dormitories to be constructed on this site with building's cardinally orthogonally aligned. The building considered in this example is the "East Dormitory." It is the one to be placed parallel to South Wright Street.

The 2016 East Dormitory Architectural Model by Prairie Sky Consulting in association with Bond Bryan Architects is licensed under a Creative Commons Attribution-ShareAlike 4.0 International License [Creative 2016]. Under the terms of this license a user of this information may "copy and redistribute" and "transform and build upon" the models for any purpose, including commercial use. If changes are made and redistributed, the user is required to share their changes under the same license, reference the original source documents, and explicitly identify the changes made.

The files available to support the evaluation of the Early Design stage testing described in this book include [East 2016]: simulated design drawings in .pdf and .dwg formats, Coordination and COBie STEP Files, COBie Spreadsheet File, and Quality Control (QC) report files. An example image from the project is shown in Figure 5.

Verification

Given a control data set, the next step in testing is to decide what criteria will be used to determine the quality of the information provided. The first of two steps in the answer to this question is called "verification." Verification answers the question, "Is the information provided in the correct format?" This directly leads to the question of the possible formats available, and rules for testing these formats.

The reason that at least one person in each company understands the COBie standard is that the requirements for COBie verification and the associated testing procedures are part of the standard itself. The following COBie testing rules are defined on page 221 of the standard in a Section called "4.2.8.1.2 Quality control test rule definition" [NIBS 2014].

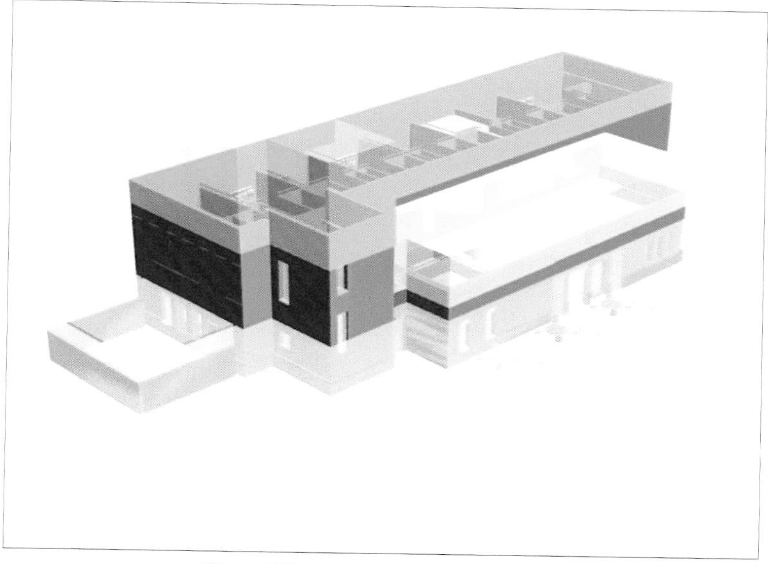

Figure 5 East Dormitory Test Building

- "AtLeastOneRowPresent" - at least one row present
- "CrossReference" - referenced Worksheet Name column
- "NotNull" - text value that is not NULL or empty
- "NotEmpty" - text value or NULL
- "OneAndOnlyOneFacilityFound" - one Facility row allowed
- "Unique" - unique within the scope of specific worksheet
- "ValidNumber" - a valid number, NULL is not acceptable
- "ValidNumberOrNA" - valid number, or NULL
- "ZeroOrGreaterOrNA" - valid number greater than zero, or NULL

The application of each of these testing rules to each column of data in a COBie File is clearly defined in the standard. The COBie standard also identifies differences in the application, or "coverage," of these rules for COBie design and construction deliverables.

Given the need for humans to evaluate large COBie files, a series of automated testing tools were developed that reduce the need for line-by-line evaluation of complex COBie data. The COBie verification testing tool of record is an open-source software tool that works in stand-alone mode or with a open-source building information modeling server. At least one person in each design office producing COBie should download and learn to use these tools to perform internal Quality Control checks on COBie files. These tools (at the time of this book's publication) are:

- COBie Plugin for bimServer.org [Bogen 2015a]. Quick Start Guide: [Bogen and East]
- COBie QC Reporter Command Line Tool [Bogen 2016]

The COBie Plugin is used when testing SPFF Files. The COBie QC Reporter is used when testing COBie Spreadsheet Files. The reports provided by each tool are the same.

Figure 6 provides a picture of a portion of the report produced by the QC tools. The summary table, in Figure 6, provides a row for every data tab within a COBie spreadsheet. Note that COBie.Instruction and COBie.PickList are never checked. The second column of the report identifies the number of rows found for the give COBie Tab. There is one row provided for COBie.Facility. The single row of in COBie.Facility is correct since a COBie File may only contain one building.

The next two columns of Figure 6 show where warnings and errors would be reported. The count of warnings and errors are based on the number of rows of data that violate the criteria defined in NBIMS-US V3. If a data row has more than one warning or error, the row will only be counted once. If there is a non-zero value in these columns, then a hyperlink is provided that lists the specific warning or error and identifies each row of COBie File data that should be reviewed or corrected. Once the complete set of Error Counts for all rows equals zero, then the format of the COBie file can be said to be verified against NBIMS-US V3 requirements. The result shown in Figure 6, is an example of just such a verified file.

COBie QC report - Design Deliverable

2016-03-11T09:15:36

Instructions

Save a new copy of your COBie spreadsheet to a new file. The errors in this report will match by Worksheet Name, Row Number, and object name to the data in your file. For example, a problem with the fifth space, named "Vestibule," willbe listed as Space[5]:Vestibule. Once each relevant issue is fixed, re-run your file.

1 COBie Summary Table

Note: Row/Column checks are only listed for worksheets containing at least one row, but row count checks are listed for Contact, Facility, Space, Type, and Component (worksheets that must contain data) regardless of row count.

Worksheet	Has Rows?	Row Count	Row Warning Count	Row Error Count
Contact	true	[illegible]	0	[illegible]
Facility	true	[illegible]	0	[illegible]
Floor	true	[illegible]	0	[illegible]
Space	true	[illegible]	0	[illegible]
Zone	true	[illegible]	0	[illegible]
Type	true	[illegible]	0	[illegible]
Component	true	[illegible]	0	[illegible]
System	false	[illegible]	n/a	n/a
Assembly	false	[illegible]	n/a	n/a
Connection	false	[illegible]	n/a	n/a
Spare	false	[illegible]	n/a	n/a
Resource	false	[illegible]	n/a	n/a
Job	false	[illegible]	n/a	n/a
Document	true	[illegible]	0	[illegible]
Attribute	true	[illegible]	0	[illegible]

Figure 6 QC Report Example

It is interesting to note that in an updated version of the COBie data for this project published subsequent to the preparation of this book, one new class of errors was encountered. Given the unusual nature of this error type, it bears discussion. Spreadsheet users will already know that the information presented on screen or print copy may not be the same as the information stored in the spreadsheet. For example, date information in a spreadsheet is saved as the number of seconds from a base date. We see the information as a human readable date, but it is stored as a long integer in the spreadsheet. The new type of problem encountered was of a similar nature. The value appearing in certain spreadsheet cells was the result of a formula. That formula was identified as an error in the QC report, but a visual inspection appeared that a correct value was provided. If there is an error in the file, and the values appear correct, users should check to make sure that what is actually stored in the cell is the needed value, and not an auto generated value or formula.

Another often encountered problem where what is presented differs from what is stored in the spreadsheet is the case of non-printing ASCII characters. Such characters often find their way into COBie spreadsheets when users manually cut-and-paste needed information from an external source, such as word processing document or PDF file into a COBie spreadsheet.

Validation

It is not enough to know that the data provided is in the correct format. Validation answers the question, "Is the information provided what was required?" This directly leads to the question of what data is expected to be delivered in a COBie data file.

In the testing described in this book, it was assumed that the quality of COBie data should be the same as what would normally be provided by the equivalent paper deliverables at the equivalent stage of the project. The design quality criteria that "COBie data must match the drawings" has been the consistent requirement for testing since 2011.

The reason this criterion was developed is two-fold. First, without trust in the information provided, it is impossible to rely on that information. The only way to obtain trust is to show clearly and as easily as possible that the information provided is the same information used elsewhere on the project. Until this quality criterion has been met, COBie data will simply be extra work to produce a file that that no one will trust.

The second reason that COBie data must match the information provided on the equivalent, existing contract documents has to do with the concept of "standard of care." Standard of care is a legal idea meaning that what is accomplished in a given context is consistent with established professional norms. If COBie is to be successfully delivered across the widest possible set of buildings with the least possible disruption to the industry, then we should begin by requiring that COBie data be equivalent to what is already provided, the "standard of care," on existing contract deliverables. To add additional requirements adds cost, confusion, and may ultimately lead to the marginalization of COBie to high-end projects that are able to absorb additional overhead costs.

Given the diversity of building types and design practices, it is not possible to develop objective criteria for the quality of data in a COBie deliverables for any possible project. The practical answer to the question of COBie data quality during design is simply that "COBie data must match the drawings." Given that design deliverables must meet the commonly accepted standard of care for the specific building being built, COBie simply follows that current standard of care. COBie leaves it to the software to correctly report out "data that matches the drawings." In following each project's standard of care COBie rejects an *a priori* context-free data quality specification such as Level of Development or Level of Detail. Should, however, such requirements be present, the COBie data structure can certainly accommodate any non-geometric data that is required.

The COBie schema was designed to capture information about managed building assets. These are products and equipment with moving parts that have to be regularly maintained. This is often described by the phrase, "COBie contains information about Pump-5 in Room-3." If we are considering, as we are in this book, where such information at the early design stage is found on existing contract documents, then the answer is clear. COBie data is simply the combined set of information found on drawing schedules.

To validate the building information delivered in COBie, the test model has been developed so that both COBie data and drawings are provided. Manually checking that COBie data information matches that found in the design schedules validates the quality of the delivered files.

APPROACH

Most COBie requirements already exist as part of standard design and construction processes. The primary source of COBie design data is drawing schedules. At its core, COBie is simply the combined set of all drawing schedules. As a result, most of the data in COBie is already familiar to anyone designing or building buildings.

For example, designers allocate a unique number and description for each building space. The list of all the spaces is found in the Room Finish Schedule. That exact same data is required by COBie. The unique name and description are captured in a COBie Tab called "Space". The room finishes themselves are contained in a COBie Tab called "Attributes".

The products installed in the building are also represented in schedules. Some schedules, such as door schedules, list each product. Other schedules, such as lighting schedules, list only the types of products. Individual products of that type are identified by a "mark" on a drawing. That exact same data is found in COBie. Each individual product in a building is captured in a COBie Tab called "Component". The overall product types are captured in a COBie Tab called "Type". The properties of each of the products themselves are contain in the COBie Tab called "Attributes".

Since COBie data is already being used in the design and construction process, the job of software systems is simply to make that data be available both for printing and for exporting to COBie. The printed version of the design data is used to visually communicate with our team members and stakeholders. The COBie version of the same design data is used to streamline the delivery of that information to the systems used by those same team members and stakeholders.

To reinforce the link between existing paper contract documents and COBie data requirements, the most important design quality criteria for COBie submissions is that the data in the COBie file match that provided on the design drawings.

While the majority of COBie data is already present in the design drawings (and associated data) in every building project, there are a few additional pieces of COBie data that help assist owners manage their buildings. For example, to ensure that COBie data coming from multiple buildings can be compared, COBie requires that "categories" be added to schedule data.

By requiring specific categories across a campus or enterprise, an owner will be able to find all spaces within a campus or portfolio that support specific types of activities. A key purpose of this book is to help the reader identify non-standard additions to design schedules, where they occur, and how to embedded that data into standard approaches. Once accomplished, the COBie data fields will not be thought of as an extra process, but something produced as standard on all projects.

To ensure that COBie data is consistently captured, most designers will find it helpful to setup an ARCHICAD template to ensure that some small COBie detail isn't missed during the design. A well-organized template means that the process of delivering COBie can be as smooth as possible. This approach to COBie can be merged into standard office templates.

Developing a robust template for COBie does take a significant amount of time for one person in a company but that one-time, up-front commitment can mean that COBie simply becomes "business as usual." As in most other endeavors, it's easier to produce COBie correctly from the start, than to debug a COBie file and have to come back and fix it after the design.

The following sections explain the tips and tricks for embedding a COBie approach within a typical early design process using ARCHICAD. This information is organized in order of how the data is displayed in a COBie spreadsheet.

ARCHICAD AND COBIE

The approach to delivering COBie is described in the following sections as a step-by-step process according to the order of information presented in a COBie spreadsheet. The first section begins with a description of the COBie.Instruction Sheet. Next, information about the Project, Site, and Building are provided. Following those sections, spatial information pertaining to the Floors, Spaces and Zones of a building are described. Alongside this, the designer develops the building elements (known as Components, Types and Systems in COBie). As data is added to each of these parts of COBie, data automatically populates the Attribute tab to fill in the information that appears on drawing schedules.

Before starting, the reader is encouraged to download and review the dormitory building used as an example to demonstrate this process [East 2016]. A full description of this building, and available test files, were provided previously in this book.

COBIE.INSTRUCTION

Overview

The COBie.Instruction Tab contains general information about the COBie file. This information is primarily of interest to people viewing a COBie spreadsheet.

When COBie data is automatically generated by software, the information provided is based on the templates used in that software. The contents of COBie Instruction are not specified by NBIMS-U V3. As a result, there are no Quality Control requirements for COBie.Instruction. As of the time this book was prepared, there were no documented uses of information on the COBie.Instruction Tab being used for commercial software import. Those who do export COBie.Instruction typically do so from a standard template for each project.

An aspect of the COBie.Instruction Tab useful to people viewing a COBie File is the explanation of how color coding can be used on later COBie Tabs. NBIM-US V3 does not require color coding but allows such coding, as shown in Figure 7, to be used to simplify human review of the information provided.

In the many COBie Spreadsheets delivered today, color coding may not correctly reflect the specific requirements of a given COBie deliverable at a specific phase of the project. This is because color coding is not identified as part of NBIMS-US V3. Colorizing COBie Columns should be accomplished based on the specifics of the regional, owner, contract requirements specifying individual COBie deliverables.

Color Code	Definition
Yellow	Data required at this stage.
Salmon	Required data provided by another COBie Tab
Purple	Data required, if system generated (else "n/a")
Green	Data required, if specified (else "n/a")
White	Data not required at this stage (expected value of "n/a")
Grey	Informational (COBie.Instruction & COBie Tabs Row 1 only)
Blue	Extended data fields (may not be supported)

Figure 7 COBie Color Coding

Process

No information is required to be input by an ARCHICAD user.

COBIE.CONTACT

Overview

The COBie.Contact Tab includes all companies involved in the project who are contributing to the COBie information. It is, in simplistic terms, a project directory for all parties to a project.

Process

COBie.Contact information can be setup in the ARCHICAD Project Info Dialog Box. This is located by navigating the following path: File > Info > Project Info. The completed dialog box is shown in Figure 8.

For the East Dormitory Project, note that additional title block data fields are filled-in under Contact Details. These fields include: Contact Fax, Contact Web, Contact Twitter and Contact Office Email. In addition, the Client Company field, under Client Details, has also been filled-in.

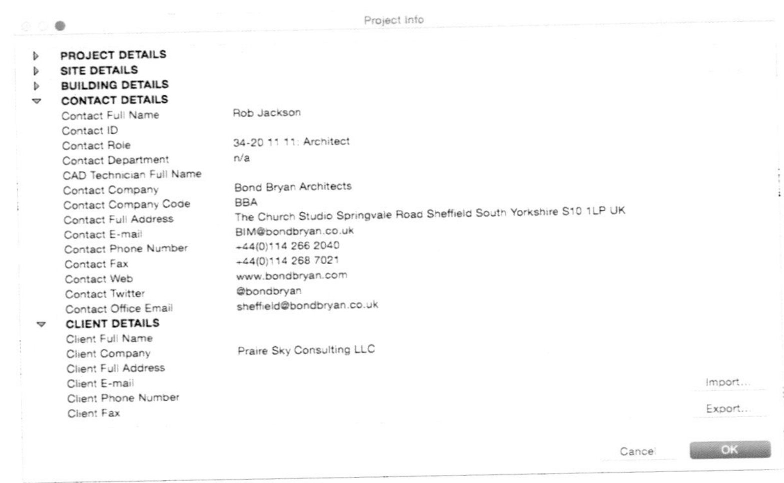

Figure 8 Project Info Dialog for COBie.Contact

Mapping

Figure 9 provides the mapping between the data found in the Project Information Dialog Box and the COBie.Contact Tab. These fields are displayed as they appear in ARCHICAD's Project Info Dialog Box.

Best Practices

To comply with increasing restrictions on the exchange of personally identifiable information, most projects do not allow the listing of individual people in COBie.Contact. This information was initially provided in the sample project to demonstrate the mapping, but later removed prior to final publication of the COBie Dormitory project.

Sole practitioners should set up all fields in a template. Once created this information will only need to be updated when the information changes.

Companies with multiple offices can create templates for each office, or simply have an XML file that can be imported at the start of a project with the correct office details.

All Contact information from Project Info can be used as Autotext so outputs can be created using this data.

ARCHICAD Field Name	COBie Mapping
CONTACT DETAILS > Contact Full Name > Given Name	Contact.GivenName
CONTACT DETAILS > Contact Full Name > Family Name	Contact.FamilyName
CONTACT DETAILS > Contact Role	Contact.CategoryContact
CONTACT DETAILS > Contact Department	Contact.Department
CONTACT DETAILS > Contact Company	Contact.Company
CONTACT DETAILS > Contact Company Code	Contact.OrganizationCode
CONTACT DETAILS > Contact Full Address > Address	Contact.Street
CONTACT DETAILS > Contact Full Address > Postal Box	Contact.PostalBox
CONTACT DETAILS > Contact Full Address > City	Contact.Town
CONTACT DETAILS > Contact Full Address > State/Province	Contact.StateRegion
CONTACT DETAILS > Contact Full Address > Postcode/ZIP	Contact.PostalCode
CONTACT DETAILS > Contact Full Address > Country	Contact.Country
CONTACT DETAILS > Contact E-mail	Contact.Email and .CreatedBy
CONTACT DETAILS > Contact Phone Number	Contact.Phone

Figure 9 ARCHICAD Project Info to COBie.Contact

Impact

While the name of and address of the company designing the building will be found on a drawing title block, that information is not, strictly speaking, required by a model authoring tool. In terms of data that was typically input into Project Info before needing to produce COBie, there are probably only two (2) fields of data that were not input before. The first is the Email address of the company. The second is the Category for the office. Since these will be the same for any specific design office, creating a preset template automates this work completely.

Given that the COBie required information is found on drawing layout sheets for traditional deliverables, it makes sense to tackle this data directly within ARCHICAD. Figure 10 provides an example of where COBie.Contact information is displayed on drawing sheets.

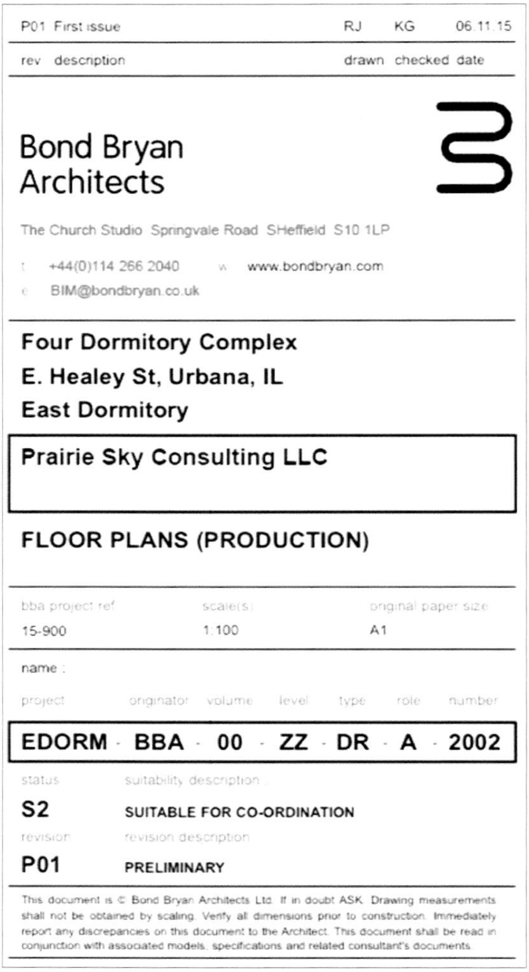

P01 First issue	RJ	KG	06.11.15
rev description	drawn	checked	date

Bond Bryan
Architects

The Church Studio Springvale Road SHeffield S10 1LP

t +44(0)114 266 2040 w www.bondbryan.com
e BIM@bondbryan.co.uk

Four Dormitory Complex
E. Healey St, Urbana, IL
East Dormitory

Prairie Sky Consulting LLC

FLOOR PLANS (PRODUCTION)

bba project ref	scale(s)	original paper size
15-900	1:100	A1

name :

project	originator	volume	level	type	role	number
EDORM	**BBA**	**00**	**ZZ**	**DR**	**A**	**2002**

status	suitability description
S2	**SUITABLE FOR CO-ORDINATION**

revision	revision description
P01	**PRELIMINARY**

This document is © Bond Bryan Architects Ltd. If in doubt ASK. Drawing measurements shall not be obtained by scaling. Verify all dimensions prior to construction. Immediately report any discrepancies on this document to the Architect. This document shall be read in conjunction with associated models, specifications and related consultant's documents.

Figure 10 COBie.Contact on Drawing Title Block

COBIE.FACILITY

Overview

COBie.Facility information is comprised of three different types of information: Project, Site and Building.

Project (IfcProject), Site (IfcSite) and Building (IfcBuilding) form a core approach to any IFC model with Floors (IfcBuildingStorey), Spaces

(IfcSpaces) and Components (IfcElements) all related to this upper level structure.

Process

The ARCHICAD dialogue in Project Info was amended by GRAPHISOFT in version 18. The rearrangement of the dialogue box was designed to provide more correlation between the Project Info dialogue and the IFC Manager, which in turn is also how COBie is organized.

As a result, almost all of the COBie.Facility information can be completed in Project Info (in ARCHICAD 18 and above). The only exception is the COBie.Facility.Category field (IfcClassificationReference) which is completed in the IFC Manager. This information can be setup at the beginning of a project when a project template is created.

The dialog box is shown in Figure 11. For the East Dormitory project, note that additional values Project Code and Project Number have also been completed in the dialogue box for other project requirements not required by COBie.

This same information may also be viewed in ARCHICAD's IFC Manager, as shown in Figure 12. Additional values are included by the model author purposes other than COBie. These are all contained under a Property Set entitled 'BBA_Pset_Template' and include BBATemplateAuthor, BBATemplateCreationDate and BBATemplateVersion.

Drilling down to the next level of the IFC tree-structure, information site-related information can be seen, as shown in Figure 13. Note that several additional values are included to support requirements of BS 1192-4. These COBie-UK required fields are: RefLatitude, RefLongitude, RefElevation and LandTitleNumber.

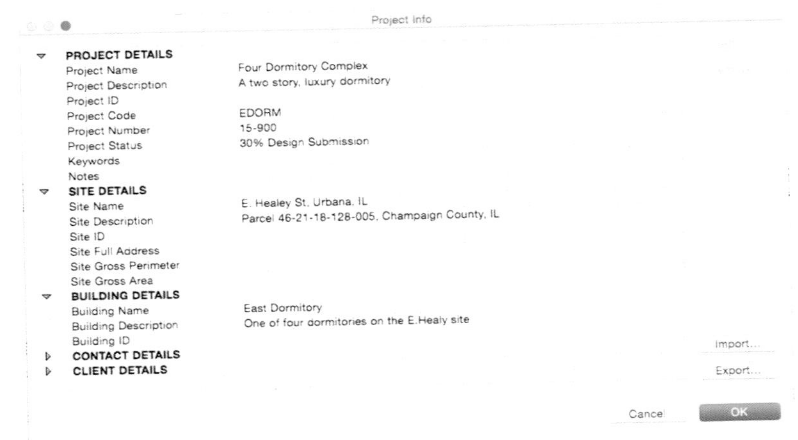

Figure 11 Project Info Dialog Box for COBie.Facility

Figure 12 IFC Manager for Project

Note: Figures 12 and 13 show "Show only Scheme items" (top right button of dialogue). Further standard properties are available within ARCHICAD out-of-the-box.

Notice in Figure 13 that the fields RefLatitude, RefLongitude and RefElevation are greyed-out. These values are controlled by values entered by navigating to Options > Project Preferences > Project Location. RefLatitude is controlled by Latitude. RefLongitude is controlled by Longitude. RefElevation is controlled by Altitude. Information for the sample project can be seen in Figure 14. This information is not required for typical US COBie usage.

IFC Manager

All Selected: 1 Editable: 1

- Four Dormitory Complex
  - E. Healey St, Urbana, IL
    - East Dormitory
      - 3. Roof
      - 2. Level Two
      - 1. Level One
      - 0. Foundation

Name	Value	Type
Attributes		
Name	E. Healey St. Urbana, IL	IfcLabel
Description	Parcel 46-21-18-128-005, Champaign County, IL	IfcText
ObjectType	Site	IfcLabel
CompositionType	ELEMENT	IfcElementCompositionEnum
LandTitleNumber	n/a	IfcLabel

IFC Groups
IFC Zones
IFC Systems
Actors
Space Occupants
Time Series Schedules

New Property / Classification... Apply Predefined Rule

Figure 13 IFC Manager for Site

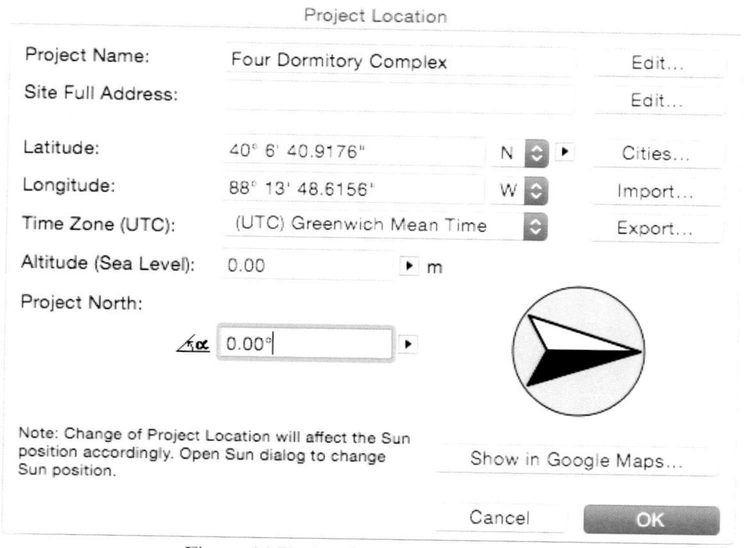

Project Location

Project Name:	Four Dormitory Complex	Edit...
Site Full Address:		Edit...
Latitude:	40° 6' 40.9176" N	Cities...
Longitude:	88° 13' 48.6156" W	Import...
Time Zone (UTC):	(UTC) Greenwich Mean Time	Export...
Altitude (Sea Level):	0.00 m	
Project North:	0.00°	

Note: Change of Project Location will affect the Sun position accordingly. Open Sun dialog to change Sun position.

Show in Google Maps...

Cancel OK

Figure 14 Project Location Information

Once values have been completed in the Project Info dialogue and the user opens the IFC Manager (Window > Palettes > IFC Manager), then values previously inputted are automatically mapped into the IFC tree structure in Figure 15.

Note that several additional values are included to support requirements of BS 1192-4. These are identified under the COBie.BuildingCommon_UK and Pset_BuildingCommon property sets. The forms shown in Figure 15 provide the data required to comply with

UK requirements where employer's information requirements dictate. This information is not required for typical US COBie usage.

> **Note**: Figure 15 shows "Show only Scheme items" (top right button of dialogue). Further standard properties are available within ARCHICAD out-of-the-box.

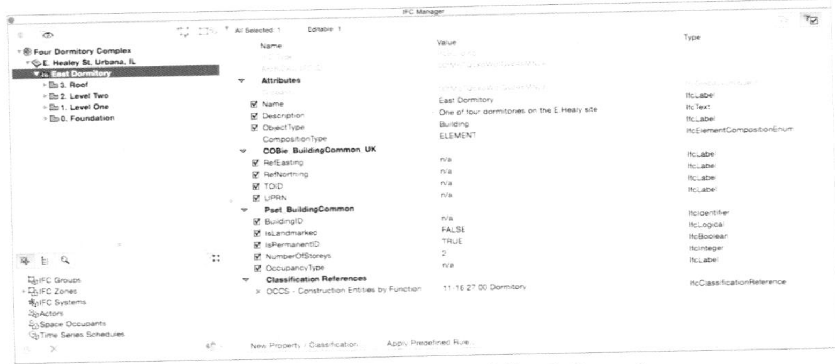

Figure 15 IFC Manager for Building Information

The final part of the process is to apply a Classification, called a Category in COBie, to the Building level. This is done by navigating to the third tier down in the tree structure on the left hand side in the IFC Manager (below Project and Site levels). The user then needs to 'Apply Predefined Rule…' at the bottom of the right hand window. Select the required table and then the required value appropriate to the building.

The choice of Classification will vary dependent on the country and specific client requirements. Large owners, for example, may have their own established facility category codes. Ideally, the specific Facility.Category code required will be identified by the client's COBie Guide (in the US) or Employer Information Requirements (EIR) (in the UK). As with the other instances where classification of COBie data is required, the required value for Facility.Category should be clarified by the Architect and provided as requested by the Owner.

If an owner does not know the answer to the question, then the default OmniClass should be used in the US. The default selection for Facility Category can be found in US Table OmniClass Table 11 (Construction Entities by Function). The default UK Facility Category for COBie-UK-2012 is to use Table D (Facilities) from Uniclass. The UK is currently in transition for Classification and in the near future is likely to move to Table En (Entities) for Building Classification from an updated Uniclass system aligned with ISO12006-1:2015.

Requirement

In all, a user is only required to complete a maximum of 8 fields of data to satisfy all the Facility data requirements of COBie. Below, these requirements are broken down into Project, Site and Building:

Project Information:

The following Project fields are required as a minimum:

- Project Name
- Phase (referred to as Project Status in ARCHICAD)
- Project Description (if specified by the client)

For projects using COBie-UK-2012, the value for Phase must equal one of the following values:

- CIC 1 Brief
- CIC 2 Concept
- CIC 3 Design Development
- CIC 4 Production Information
- CIC 5 Constructed Information
- CIC 6 Handover
- CIC 6A Post fit-out Handover
- CIC 7 Post Practical Completion.

Site Information

The following Site fields are required as a minimum:

- Site Name
- Site Description (if specified by the client)

The following building information is required as a minimum:

- Name (Building Name)
- Category (Building Classification)
- Description (Building Description) (if specified by the client)

Mapping

The following needs to be completed in the Project Info dialogue box to provide the required data for COBie Facility (noted in order as it appears in ARCHICAD's Project Info):

Figure 16 provides the mapping between the data found in the Project Information Dialog Box and the COBie.Facility Tab. These fields are displayed as they appear in ARCHICAD's Project Info Dialog Box. The remaining fields required by COBie.Facility, such as Facility.Catgory, and the additional fields required by COBie-UK are provided directly through the IFC Information Manager.

ARCHICAD Field Name	COBie Mapping
PROJECT DETAILS > Project Name	Facility.ProjectName
PROJECT DETAILS > Project Description	Facility.ProjectDescription
PROJECT DETAILS > Project Status	Facility.Phase
SITE DETAILS > Site Name	Facility.SiteName
SITE DETAILS > Site Description	Facility.SiteDescription
BUILDING DETAILS > Building Name	Facility.Name
BUILDING DETAILS > Building Description	Facility.Description

Figure 16 Project Info to COBie.Facility

Best Practices

Consider using Project Name, Site Name and Building Name fields on layout sheets. Figure 17 demonstrates where COBie data would appear in a typical drawing title block. Ensuring that COBie data matches the drawings is a way to improve the quality of the COBie deliverable and a way to ensure users who issuing traditional deliverables remember to provide this data. As noted with COBie.Contact, all Project Info can be used as Autotext so outputs can be created using this data.

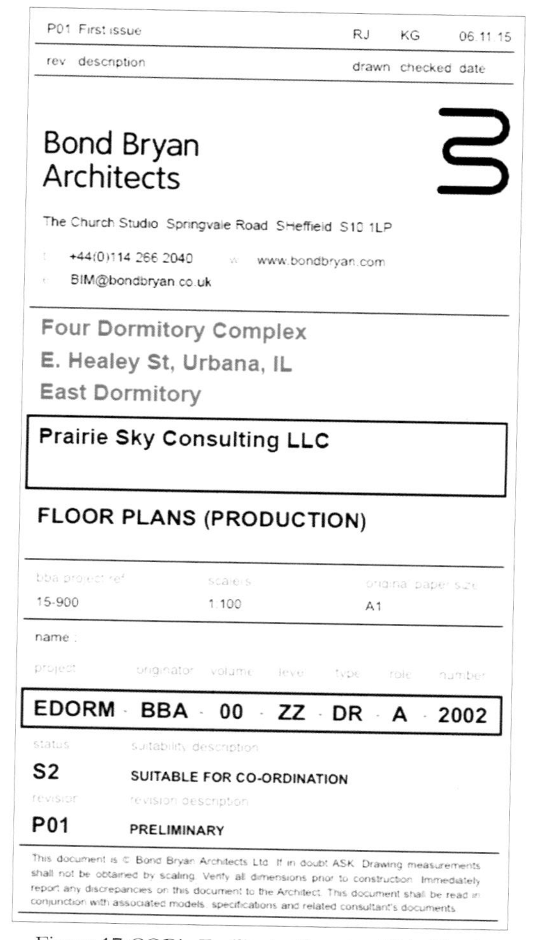

Figure 17 COBie.Facility in Drawing Title Block

Impact

The concept of a Project, Site and Building are fairly clear to understand for all Architects. Many users will have been using the Project Info dialogue without even realizing they have effectively been completing COBie information.

For COBie use in the US, there are only 8 pieces of data to complete to provide all the Facility data. The only one that requires a little more training and understanding is the application of the Classification

Reference (Category. COBie-UK (BS1192-4:2014) identifies additional information that can be included beyond standard COBie specification.

COBIE.FLOOR

Overview

Floors are fairly self-explanatory and to many ARCHICAD users who have been modeling for even a short time they will understand the need to manage Floors or what may be referred to as Stories. In IFC a Floor is referred to as an IfcBuildingStorey.

Process

The basic information required for COBie.Floor is provided directly in ARCHICAD Design > Story Settings. This information includes Name, Elevation, and Height-to-Next, as shown in Figure 18.

Figure 18 Story Setting Dialog

The Name and Elevation fields are automatically mapped into the IFC Manager. The Height-to-Next data is exported from ARCHICAD but not shown in the IFC Manager.

Additional COBie data is required to be provided through the IFC Manager. Figure 19 provides the dialog box to update the Floor Description. Figure 20 illustrates the dialog to provide the Floor.Category.

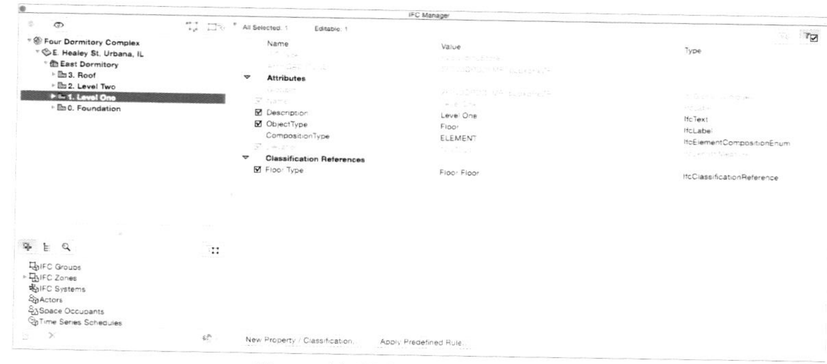

Figure 19 IFC Manager for Floor

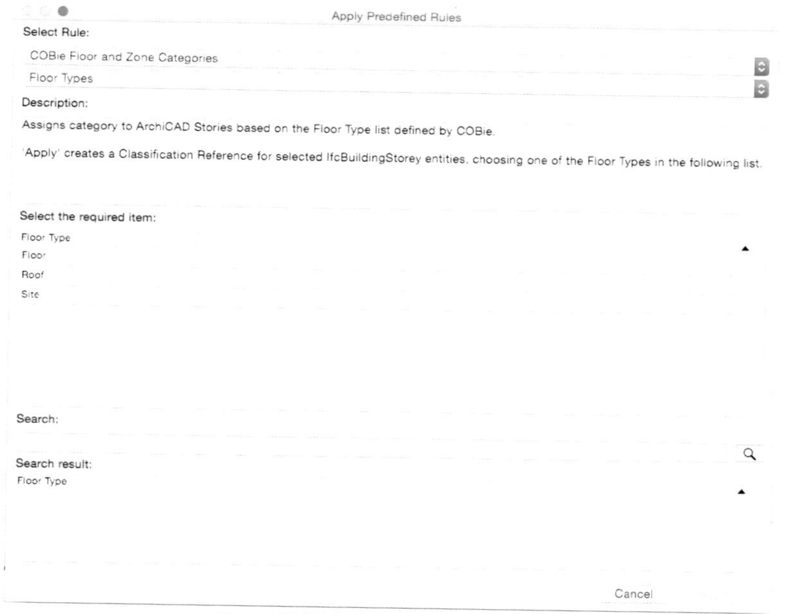

Figure 20 IFC Classification for Floor

Requirements

The following Floor fields are required as a minimum:

- Name
- Category

The following Floor fields are required if specified by the client:

- Description
- Elevation
- Height (Height to Next)

The following Floor fields are automated by ARCHICAD:

- CreatedOn
- ExtSystem
- ExtObject
- ExtIdentifier

The following Floor fields are produced by completing the required Contact information:

- CreatedBy

Best Practices

Presetting templates preloaded with Description and Classification References (Category) means users can simply remove the Floors they don't need in the Story Settings and then adjust the Elevation and Height to Next (Height) accordingly. Users only need to complete these details manually where they are adding new Floors (Building Stories) to a project.

Impact

ARCHICAD users creating any model will already be completing Name, Elevation and Height as part of a standard approach to setting up a model, so only applying the Description and Category (IfcClassificationReference) is additional. As mentioned in the tips and tricks, this can be preset for most applications and therefore there is little extra work in creating Floor information for ARCHICAD users.

COBIE.SPACE

Overview

Spaces can sometimes be referred to as Rooms although Spaces is a more generic term to cover 'Rooms' without physical boundaries such as external Spaces or other open plan areas. In addition, there may be large areas with shared set of walls that provide multiple functions. Each of these sub-areas with may have different finishes, or types of components. Spaces are defined in COBie as physical areas that share a common function. A core concept in IFC (and therefore COBie) is to have spatial relationships between Spaces and Components in a Building. All Spaces must be modelled.

In simple terms, the COBie Space sheet should be thought of as a Schedule of Accommodation. This list can be used to confirm the client's requirements at the early design stages. The creation of these Spaces also allows other information to be added to these Spaces to create further architectural information such as Room Data Sheets.

Since buildings often have exterior and roof mounted equipment spaces must also be developed for those areas outside the building. Figure 21 shows the spaces of the example East Dormitory project that cover areas outside the building envelope.

Process

To create Spaces for COBie, ARCHICAD's Zone Stamp must be utilized. The Zone Stamp is automatically classified as an IfcSpace. Note that ARCHICAD's Zone Stamp should not be confused by IfcZones (Zones in COBie). These two things are different.

For the data, the Name (Zone Stamp No.) and Description (Zone Stamp Name) are completed directly in ARCHICAD's Zone tool. These two data fields are automatically mapped into the relevant IFC fields (i.e. Zone Stamp No. maps to Name and Zone Stamp Name maps to LongName).

Spaces can be defined as volumes with the height of the space multiplied by floor area of the space. Both of these dimensions are surprisingly more complex than one might initially think. The following paragraphs describe some of the nuance needed to correctly model spaces.

The first issue to consider is that of the height of the space. The differences in the definition of spaces in IFC (and COBie) and ARCHICAD are illustrated in Figure 22 [bSi 2013]. Referring to Figure 22, an ifcSpace extends from the "FinishFloorHeight" to the "Height" or the bottom of the slab above. An additional measure, "FinishCeilingHeight," separates the ifcSpace (if needed) into conditioned space below and interstitial space above.

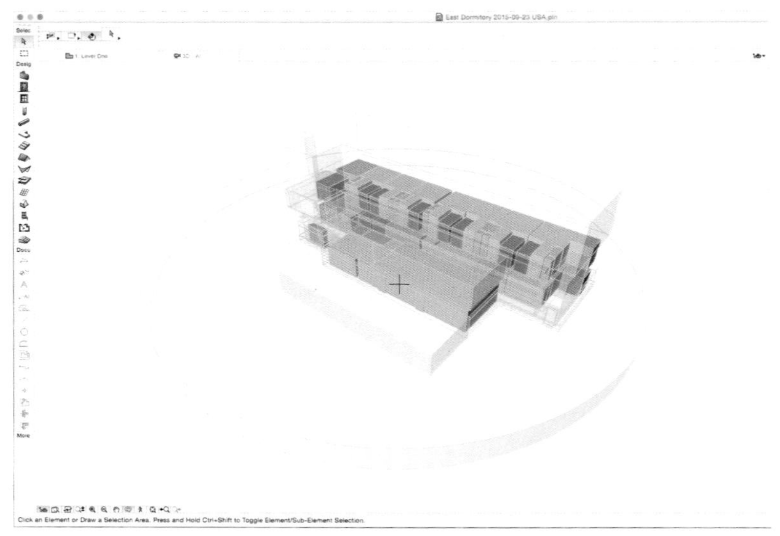

Figure 21 Example Project Exterior Spaces

In COBie, identification of the "spatial containment" is required for every piece of equipment or product installed in the building. Identification of the interstitial space is critical since equipment placed within that space will most likely be accessed, for purposes of operations and maintenance, from the conditioned space below.

In contrast to the required IFC definition of space height ARCHICAD spaces are created based on the "FinishFloorHeight" to the "FinishCeilingHeight". Products and equipment above the ceiling are not identified in ARCHICAD as being within the space below. This means that the associated equipment found in that ceiling may not be correctly exported.

To capture ceiling mounted equipment it may be necessary to slightly increase the typical ARCHICAD ceiling elevation to capture ceiling mounted equipment. Other tools, such as Solibri Model Checker, have been used to correctly integrate engineering models that have equipment placed within interstitial spaces with the spaces below.

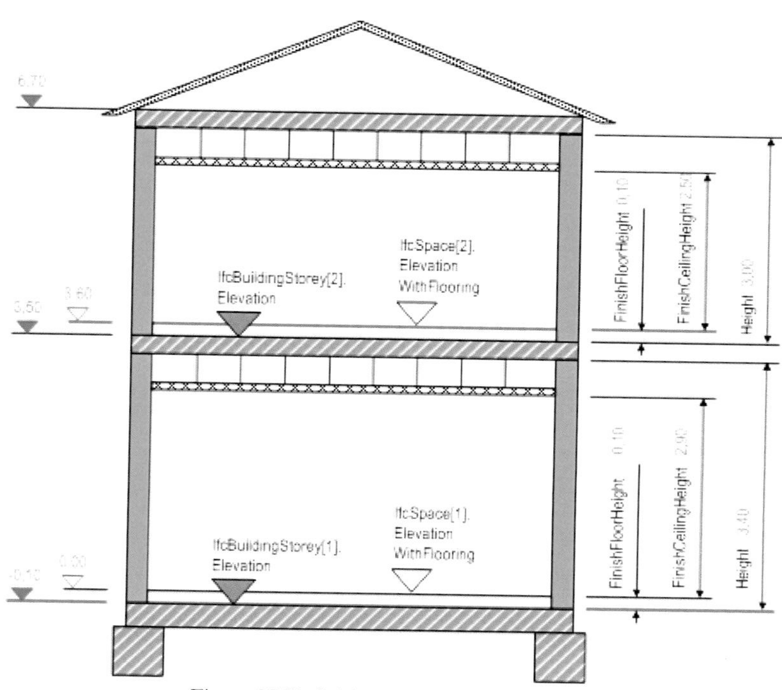

Figure 22 Definition of ifcSpace [bsi 2013]

To export Spaces (Zone Stamps), they must be switched on in the 3D window and not just in visible in the 2D window. The user can then choose "Visible" export in the translator settings. This is a common mistake made by users when exporting IFC models. (Note: It is possible to choose "entire model" as an alternative to exporting Spaces [Zone Stamps], but this may export other unwanted elements.)

As with the identification of the height of the space in IFC versus the height of the ceiling in ARCHICAD, the issue of measuring the floor area of a space is also highly complex. In 2010, the United States American National Standards Institute (ANSI) and Building Owners and Managers Association (BOMA) organizations jointly published a standard to define office space area measurements [ANSI/BOMA 2010]. A key area of interest to these organizations was to define common measures for areas within buildings that could be directly or charged to tenants, versus the areas that cannot be charged to the tenant. In addition to direct rents, issues such as charge-backs for cleaning costs and maintenance are also critical efficient building management. The resolution of such issues within COBie is that whatever method to calculate COBie.Space.GrossArea and COBie.Space.NetArea are used, it must be clearly defined in COBie.Facility.AreaMeasurementStandard.

COBie.Space data can be seen in the IFC Manager by selecting an individual space as shown in Figure 23 or by using the Zone Stamp's Element Settings dialog with Manage IFC Properties.

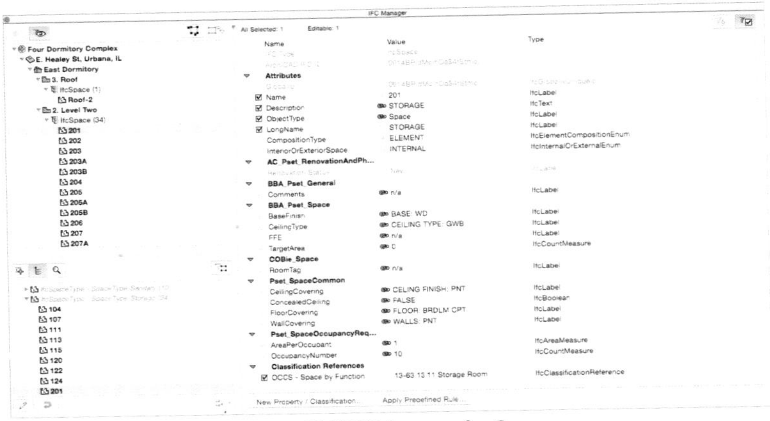

Figure 23 IFC Manager for Space

Category (IfcClassificationReference) is completed with 'Apply Predefined Rule…' applied for each Space (Zone Stamp). This can be done one-by-one for each Zone Stamp or by selecting multiple Zone Stamps that are in the same Category using manual selection or ARCHICAD's Find & Select functionality (Edit menu). The IFC Classification manager for Category is shown in Figure 24.

A user can easily and rapidly search for the required category in large Classification (Category) tables within the 'Apply Predefined Rules' dialog.

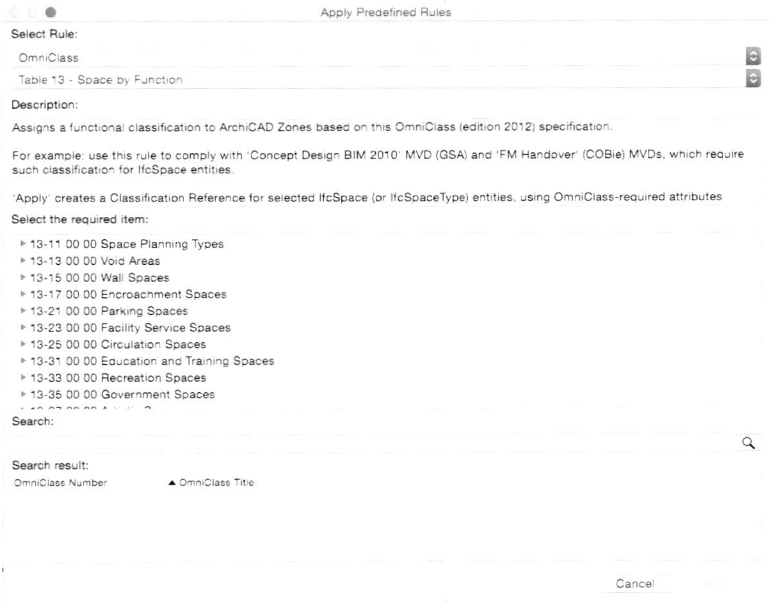

Apply Predefined Rules

Select Rule:

OmniClass

Table 13 - Space by Function

Description:

Assigns a functional classification to ArchiCAD Zones based on this OmniClass (edition 2012) specification.

For example: use this rule to comply with 'Concept Design BIM 2010' MVD (GSA) and 'FM Handover' (COBie) MVDs, which require such classification for IfcSpace entities.

'Apply' creates a Classification Reference for selected IfcSpace (or IfcSpaceType) entities, using OmniClass-required attributes

Select the required item:

- 13-11 00 00 Space Planning Types
- 13-13 00 00 Void Areas
- 13-15 00 00 Wall Spaces
- 13-17 00 00 Encroachment Spaces
- 13-21 00 00 Parking Spaces
- 13-23 00 00 Facility Service Spaces
- 13-25 00 00 Circulation Spaces
- 13-31 00 00 Education and Training Spaces
- 13-33 00 00 Recreation Spaces
- 13-35 00 00 Government Spaces

Search:

Search result:

OmniClass Number ▲ OmniClass Title

Cancel

Figure 24 IFC Classification for Space

During export, COBie.Space.GrossArea is rounded to one decimal place. This is done to provide a reasonable number of significant digits in the reporting of area measurements. While many software systems contain numbers with a long string of decimal places, measuring space area to fourth or fifth digit is not really necessary. Furthermore, the delivery of such seemingly highly precise data belies actual level of accuracy needed by or available to real building users.

COBie Space.Name (the space number), COBie.Space.Description, and COBie.Space.GrossArea can be checked against either the drawing room schedule or the floor plan. Figure 25 provides another example of how COBie data is simply the data already found on design drawings.

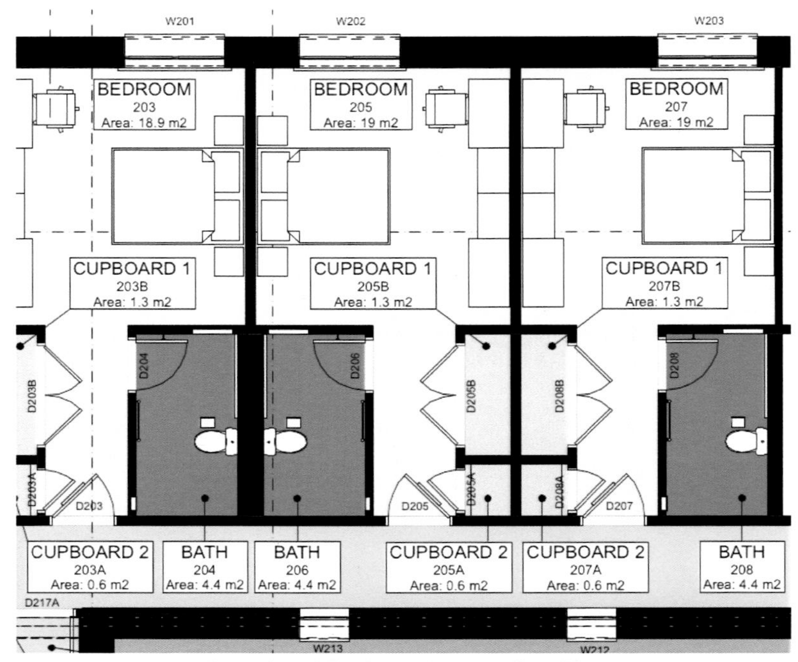

Figure 25 COBie.Space Data on Floor Plan

Requirement

The following Space fields are required as a minimum:

- Name (Zone Stamp No.)
- Description (Zone Stamp Name)
- Category (IfcClassificationReference)

At the design stages of a project, the following Space fields are required as specified by the client:

- UsableHeight
- GrossArea
- NetArea

Users at the design stage will typically not enter RoomTag data. This is because RoomTag is the location for room signage installed by the contractor.

The following Space fields are automated by ARCHICAD:

- CreatedOn
- ExtSystem
- ExtObject
- ExtIdentifier

The following Space fields are produced by completing the required Contact and Floor information respectively:

- CreatedBy
- Floor Name

Best Practices

Spaces that span multiple floors pose a specific set of issues. In general, the following best-practices apply. Elevator shafts should be modeled as a space from the bottom of the shaft to the top of the shaft. On the other hand, stairwells should be modeled as spaces on each individual floor. Modeling stairwells on every floor ensures that the areas of floor covering on the stairwell landings can be added into the room finish schedules correctly.

For scheduling purposes, it may be necessary to create additional Spaces on a hidden layer where Spaces have been modelled as double height in order to identify that elements such as Doors are connected to the Space within ARCHICAD. These "hidden spaces" are only for use within ARCHICAD and are subsequently switched-off in the COBie export.

One other point of complexity is atrium spaces where bridge Spaces penetrate that Space. While not specifically tested in this example, the preferred way to model such spaces is to have two spaces, one for the atrium and one for the bridge that is within the atrium space.

If repeat building types or regular clients are used, then Spaces information can be preset up in office templates using ARCHICAD's Favorites functionality.

Always export models from the 3D window and ensure that the Spaces (Zone Stamps) are visible before exporting. Layer Combinations combined with Views and Publisher Sets are a good way to ensure a robust process for export.

While the complexity of area measurement standards is beyond the scope of this book, it may be very helpful to the owner if areas are zoned both by function (as noted in Figure 24) and by tenant. If the tenants were "to be determined," then areas that would be "rentable" versus those that would not be "rentable" can be identified and scheduled to assist owner business decisions such as "cleaning costs and space efficiencies" [BOMA 2010].

Impact

ARCHICAD users will be familiar with using the Zone Stamp both for plan representation and to use data within Interactive Schedules (e.g. for Schedules of Accommodation). The information required by COBie will be created by even the most average of ARCHICAD users.

The main additional work compared to traditional processes is to classify the Spaces with a Category (IfcClassificationReference). For the East Dormitory project, this uses OmniClass Table 13 – Space by Function. For COBie-UK-2012 Uniclass 1.4 Table F – Spaces would be utilized. For Uniclass 2015, the equivalent table would be the SL – Spaces/ locations table. As noted previously, it is important to have the client identify the required classification system prior to starting the design.

COBIE.ZONE

Overview

Zone information is made up by allocating each Space to a Zone, with the Name (and Description) requirements for Zones are typically determined by the client at the start of a project.

It should be noted that Zone data is optional for COBie and only required where a client specifies both the need for Zones and how they wish the Zones to be split up within the Facility. As with questions of classification, if zoning requirements are not identified by the owner, a request for information should be submitted.

Requirements

The following fields are required as a minimum;

- Name
- Category

The following fields are required if specified by the client:

- Description

The following Space fields are automated by ARCHICAD:

- CreatedOn
- ExtSystem
- ExtObject
- ExtIdentifier

The following Space fields are produced by completing the required Contact and Space information respectively:

- CreatedBy
- Space.Name

Process

Zones (IFCZones) can only be created within the IFC Manager and involve dragging the required Spaces (previously created using ARCHICAD's Zone Stamp) from the top left window into the appropriate Zones created in the Assignments area in the bottom left hand window. See GRAPHISOFT's guides for more detail on this process. Figure 26 provides a screen shot of the IFC Manager for Zones.

Each Zone is allocated a Category from the list of COBie zone types:

- Circulation Zone
- Lighting Zone
- Fire Alarm Zone
- Historical Preservation Zone
- Occupancy Zone
- Ventilation Zone

During the design stage, zoning will typically be limited to Circulation Zone and/or Occupancy Zone. Note that NBIMS-US V3 provides these zone types as default example values only. As with other areas of classification, the client's input is needed. Once the required set of classifications is developed, the IFC Classification manager, Figure 27, may be used to manage the zone categories.

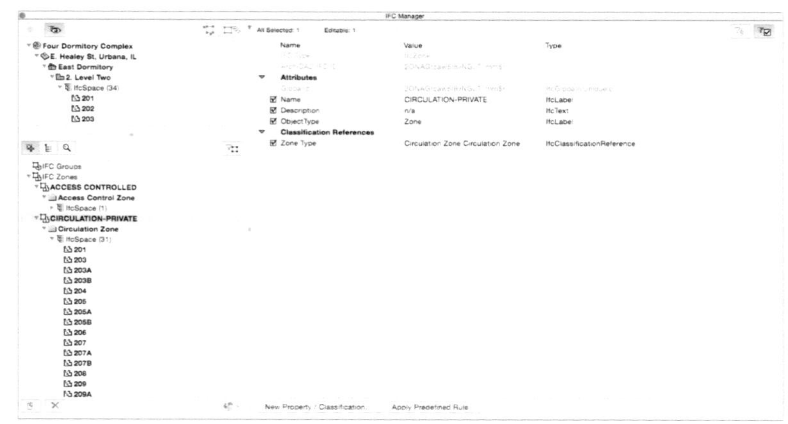

Figure 26 IFC Manager for Zone

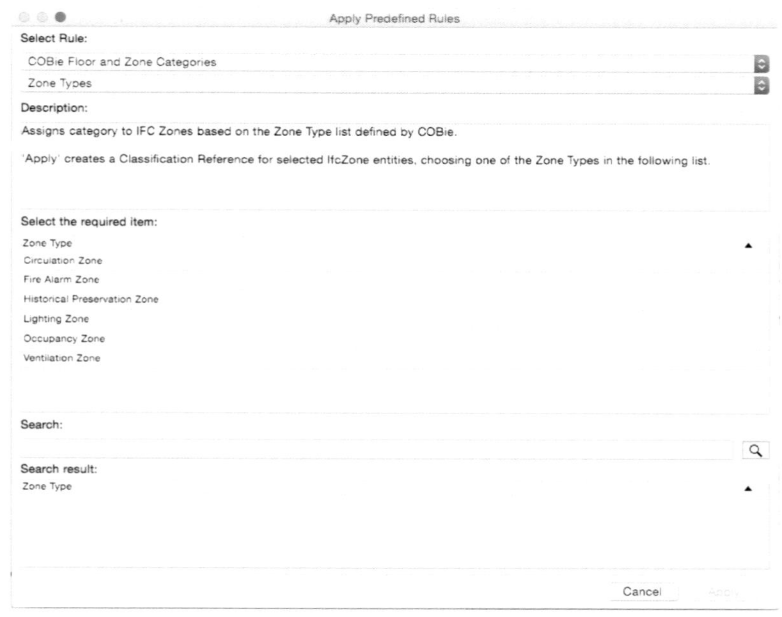

Figure 27 IFC Classification for Zone

ZONE NAME	SPACE NAME	SPACE DESCRIPTION	QUANTITY	AREA
ACCESS CONTROLLED				
	202	STORAGE	1	17.2
CIRCULATION-PRIVATE				
	201	STORAGE	1	9.0
	203	BEDROOM	1	18.9
	203A	CUPBOARD 2	1	0.6
	203B	CUPBOARD 1	1	1.3
	204	BATH	1	4.4
	205	BEDROOM	1	19.0
	205A	CUPBOARD 2	1	0.6
	205B	CUPBOARD 1	1	1.3
	206	BATH	1	4.4
	207	BEDROOM	1	19.0
	207A	CUPBOARD 2	1	0.6
	207B	CUPBOARD 1	1	1.3
	208	BATH	1	4.4
	209	BEDROOM	1	19.0
	209A	CUPBOARD 2	1	0.6
	209B	CUPBOARD 1	1	1.3
	210	BATH	1	4.4
	211	BEDROOM	1	18.4
	211A	CUPBOARD 2	1	0.6
	211B	CUPBOARD 1	1	1.3
	212	BATH	1	4.4
	213	BEDROOM	1	19.0
	213A	CUPBOARD 2	1	0.6
	213B	CUPBOARD 1	1	1.3
	214	BATH	1	4.4
	215	BEDROOM	1	24.7
	215A	CUPBOARD 2	1	0.6
	215B	CUPBOARD 1	1	1.4
	216	BATH	1	4.4
	217	HALL	1	41.5
	Roof-1	ROOF OF FIRST FLOOR	1	143.6
	Roof-2	ROOF OF SECOND FLOOR	1	357.7

Figure 28 COBie Zone as Schedule of Accommodations

By summarizing the spaces that have similar functions, COBie Zone data can be used to support the client's schedule architectural programming functions and accommodations. Figure 28 provide an example of this use of COBie Zone data.

Best Practices

While there is an issue previously identified with viewing which Spaces have been unallocated to Zones, one workflow is to set up a schedule to view the Zone Name and Space Names. This allows the user to see which Spaces have been allocated to which Zone. Unallocated Spaces will show a blank Zone Name. The user will need to return to the

IFC Manager, or use Find & Select, as described earlier, in order to allocate any missing relationships.

Impact

Many ARCHICAD users will not have traditionally used this particular workflow, however, the information is often needed by owners. The workflow can be used to schedule Spaces by departments or other splits in accommodation. This workflow is useful to provide schedules to clients showing splits of their accommodation.

COBIE.COMPONENT

Overview

With the exception of 2D, Grids, and Spaces everything placed in an ARCHICAD model, required by the COBie schema, is considered a Component in COBie terminology. ARCHICAD users often refer to them as an Element and in IFC they are considered an IfcElement. They could also be referred to as instances within a model.

When thinking about COBie, it is critical to know that COBie information is required only for products and equipment that, once the project is complete, will have regularly scheduled maintenance and repair schedules. COBie data is not required for every object in a building. The list of object types that are included and excluded is specified in the NBIMS-US V3 standard.

Process

The process of creating Components is fairly straightforward. It relies on the user modeling and placing Elements and Objects correctly within the model and ensuring that these are classified correctly with the Element Classification available under Tags & Categories. Figure 29 provides a screen-shot of the IFC Manager for Components.

More on Element Classification within ARCHICAD can be found in these series of blog posts that begin with a discussion of classification relationship [Jackson 2015a], ARCHICAD type and Library Parts [Jackson 2015b], IFC Predefined Type Attributes [Jackson 2015c]. A description of the objects and types in IFC in plain language is also available [Jackson 2015d].

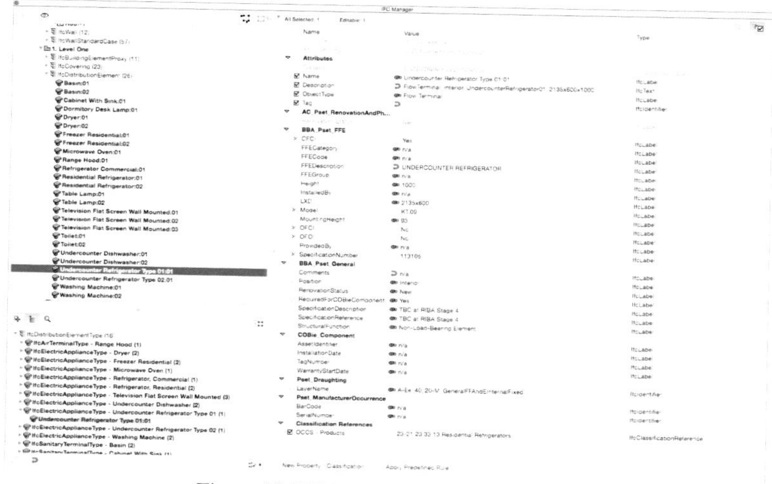

Figure 29 IFC Manager for Components

Requirements

An often misunderstood concept is that COBie data is not required for every component in a model. COBie data is only required for Components that have a preventative maintenance requirement. To identify which building are within the COBie scope, and which are not, NBIMS-US V3 Chapter 4.2, Annex A, section A.4 (page 71) explicitly identifies the IFC elements that outside the scope of the COBie specification. An important reason for at least one person in every design practice should have a copy of NBIMS-US V3 is to have access to this exclusion list.

While the referenced section within NBIMS-US V3 provides the complete set of objects to be excluded from COBie data files, the elements to be included and excluded can be summarized as described in the following three paragraphs.

First, COBie will typically include pre-manufactured, off-the-shelf building elements with model numbers and operations and maintenance schedules. These elements will usually also have some type of warranty, replacement parts lists, and consumable parts such as filters, belts, or lubricants.

Second, COBie may include engineered-to-order elements such as switchgear and elevators. During the design stage such elements are identified by larger assemblies, and that is how the elements will appear in

COBie at design. During construction, however, constituent parts may need to be individually identified because of differences in operations and maintenance requirements. In this book, whose focus is the design process, the procedure to follow to deliver complex engineered-to-order building elements in COBie.Assembly is not addressed.

Third, COBie should exclude all building element constructed on-site. Elements to be excluded, therefore, include but are not limited to structural systems such as slabs, column, beams, and walls. Products that move fluids through the building, such as pipes, ducts, and wires are also excluded from COBie.

The following fields are required as a minimum for each Component:

- Name
- Description

Each Component must have a unique Name. For example, for doors, the Name field may be D101, D102, D103 etc. as in the case of the East Dormitory project. The Description field is not prescribed although it should have a logical approach to describe what the Component in the model represents.

A very important requirement to verify that COBie data is correct is to check to see if the products designation on the drawing can be matched with the same component in the COBie file.

Following the design stage, COBie Component contains fields that allow the contractor's information to be added. Since the information will not be available to the designer, the NULL value should be provided. Per NBIMS-US V3 the null value for COBie data is "n/a". There is, however, one difference between NULL values NBIMS-US V3 and COBie-UK-2012 related to NULL values for dates. NULL dates in the UK must be entered as "1900-12-31T23:59:59" and not "n/a".

The data fields required for contractor entry are as follows:

- SerialNumber
- InstallationDate
- WarrantyStartDate
- TagNumber
- BarCode
- AssetIdentifier

Additional fields found in the IFC Manager for Components should not be entered by the designer. These fields include:

- Pset_ManufacturerOccurence.SerialNumber
- Pset_ManufacturerOccurence.BarCode
- COBie_Component.InstallationDate
- COBie_Component.WarrantyStartDate
- COBie_Component.TagNumber
- COBie_Component.AssetIdentifier

Issues

During the East Dormitory project, it was discovered that ARCHICAD Lamp objects were not recognized to be associated to Spaces when exported to COBie. A workaround was to convert these objects to Morphs, although this would mean they became static objects and could no longer have their geometry altered parametrically. This issue was reported to GRAPHISOFT and has since been resolved in later builds/versions of ARCHICAD.

Also identified as part of this project, an ARCHICAD object not used to model a specific component, such as Furniture, needed to be converted to a Morph to ensure it is associated to a Space. i.e. Modelling a Slab and classifying its Element Classification (under Tags & Categories) as Furniture will result in the Component not being associated to a Space. This was reported to GRAPHISOFT and has since been resolved.

The following objects' default setting is to assume that a sink is included in the placed object (i.e IfcFlowTerminal > IfcSanitaryTerminalType with PredefinedType SINK). If the object was configured without a sink, then the user would need to manually alter the classification to Furniture (IfcFurnishingElement).

- Library Workstation 19
- Cabinet Base Corner C with Sink 19
- Cabinet Base Double Door 19
- Cabinet Base Single Door 19
- Cabinet Base Triple Door 19
- Cabinet Belfast Double Door with Sink 19
- Cabinet Belfast Single Door with Sink 19

While Element Classification for the out-of-the-box ARCHICAD objects are automated when 'ARCHICADARCHICAD Type' is utilized, the following objects have incorrect Element Classifications *(list correct at time of publication)* and must be altered manually:

- Alarm Siren 19 should be an IfcDistributionControlElement > IfcAlarmType with a PredefinedType of SIREN rather than IfcBuildingElementProxy.
- Bench 01 19, Bench 02 19, Bench 03 19 and Bench 04 19 (under Park and Street 19) should be classified as IfcFurnishingElement > IfcFurnishingElementType rather than IfcBuildingElementProxy.
- Cabinet Base Dishwasher 19 should have a PredefinedType of DISHWASHER rather than DIRECTWATERHEATER.
- Ceiling Cassette Air Conditioner 19 should be an IfcEnergyConversionDevice > IfcUnitaryEquipmentType with a PredefinedType of AIRCONDITIONINGUNIT rather than IfcBuildingElementProxy.
- Cooktops Electric 19 should be IfcFlowTerminal > IfcElectricApplianceType rather than IfcBuildingElementProxy. Cooktops Gas 19 should be IfcFlowTerminal > IfcGasTerminalType rather than IfcBuildingElementProxy.
- Dryer 19 should have a PredefinedType of TUMBLEDRYER rather than RADIANTHEATER.
- Fire Alarm 19 should be an IfcDistributionControlElement > IfcAlarmType rather than IfcBuildingElementProxy.
- Gutter Head Leader 19 should be an IfcFlowTerminal > IfcStackTerminalType with a PredefinedType of RAINWATERHOPPER rather than IfcDiscreteAccessory.
- Microwave Oven 19 should be an IfcFlowTerminal > IfcElectricApplianceType with a PredefinedType of MICROWAVE rather than IfcFurnishingElement.
- Mop Sink 19 should have a PredefinedType of SINK rather than WASHHANDBASIN.
- Range Electric 19 should be IfcFlowTerminal > IfcElectricApplianceType with a PredefinedType of ELECTRICCOOKER rather than IfcBuildingElementProxy.

- Range Gas 19 should be IfcFlowTerminal > IfcGasTerminalType with a PredefinedType of GASAPPLIANCE rather than IfcBuildingElementProxy.
- Shower Cabin 19 should have a PredefinedType of SHOWER rather than BATH.
- Shower Kit 19 should be an IfcFlowTerminal > IfcSanitaryTerminalType with PredefinedType of SHOWER rather than IfcBuildingElementProxy.
- Smoke and Heat Detector 19 should be an IfcDistributionControlElement > IfcSensorType rather than IfcBuildingElementProxy.
- Wall Mounted Air Conditioner 19 should be an IfcEnergyConversionDevice > IfcUnitaryEquipmentType with a PredefinedType of AIRCONDITIONINGUNIT rather than IfcBuildingElementProxy.
- Washer 19 should have a PredefinedType of WASHINGMACHINE rather than TUMBLEDRYER.

The ARCHICAD mapping of information into IFC fields is a powerful feature but it lacks the ability to produce a locked mapping. For example, the Description field can be mapped from geometry and core settings but it can also be easily overwritten by a user. At present there is no way to lock this in the user interface or provide any visible feedback for all Components in a model to show what has been overwritten (this could potentially be incorporated into Interactive Schedules).

Best Practices

The following are some of the tips and tricks picked up through applying COBie to the East Dormitory project and other real life implementation of COBie on projects:

1. Element Classification is key to ensuring a reliable COBie deliverable. Element Classification can be preset by using ARCHICAD's Favorites functionality.
2. For Objects, it is generally advised to leave Element Classification as 'ARCHICADARCHICAD Type' as some Objects drive the Type classification. This is particularly true of many service Components.

3. It is recommended that all Components complete the Description field as a minimum as part of any standard BIM approach. The Name field may also be used for Components not required by COBie for scheduling purposes.
4. The Description field can be fully automated using mapping of selected geometry and data in ARCHICAD's IFC Scheme.
5. Placeholder values for SerialNumber, InstallationDate, WarrantyStartDate, TagNumber, BarCode and AssetIdentifier can be setup in ARCHICAD's IFC Scheme within office ARCHICAD templates. This then requires no further action by the user in ARCHICAD.
6. All information for Components can be scheduled with Interactive Schedules. Suggest adding standard COBie Component schedules to office ARCHICAD templates.
7. Checks for duplication of Name fields can also be managed in Interactive Schedules. These are best setup again in office ARCHICAD templates.
8. Interactive Schedules can be setup to only show items required specifically for COBie. This makes understanding for users simpler.
9. Unique Names can be managed with the 'Element ID Manager' to speed up the process of Naming Components.

Impact

Modelling Components is a core process that all ARCHICAD users should be familiar with. The additional work is really in understanding the information process and how modeling will affect the output of both IFC and COBie.

COBIE.TYPE

Overview

Types are created in ARCHICAD by placing Components within the model. The Type level offers the opportunity to produce data that is common to a series of Components that are identical.

Process

Types can only be created by placing Components within the ARCHICAD model. The setup out-of-the-box template will create a certain amount of automation of Types but it is insufficient to satisfy the requirements of COBie regarding Type Names. It is therefore necessary to create complex mapping in the IFC Scheme to generate a full set of Types within a model.

Once the Types were generated for the East Dormitory project, it was then necessary to manually edit all exported Type Names and Descriptions in order to satisfy the client requirements and match the information produced for traditional 2D outputs.

Predefined Types and AssetType for Movable items were also manually edited in the IFC Manager, as shown in Figure 30.

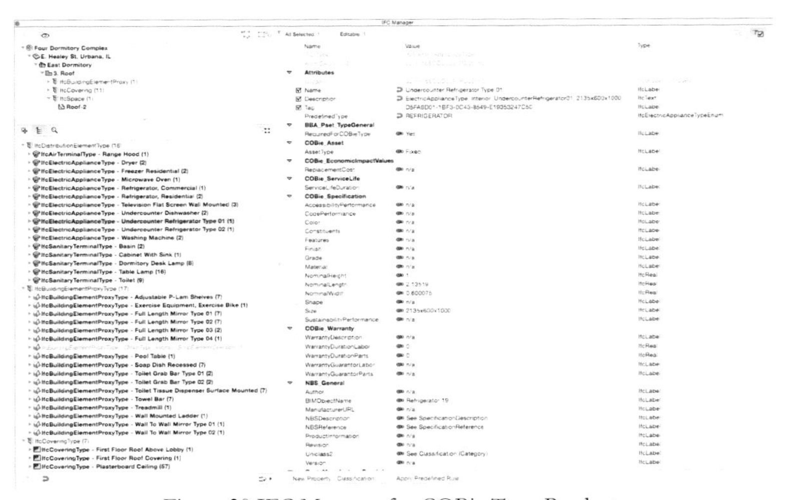

Figure 30 IFC Manager for COBie Type Products

To provide a fully compliant COBie output, it is necessary to edit all Types to apply the Category (IfcClassificationReference). As shown in Figure 31.

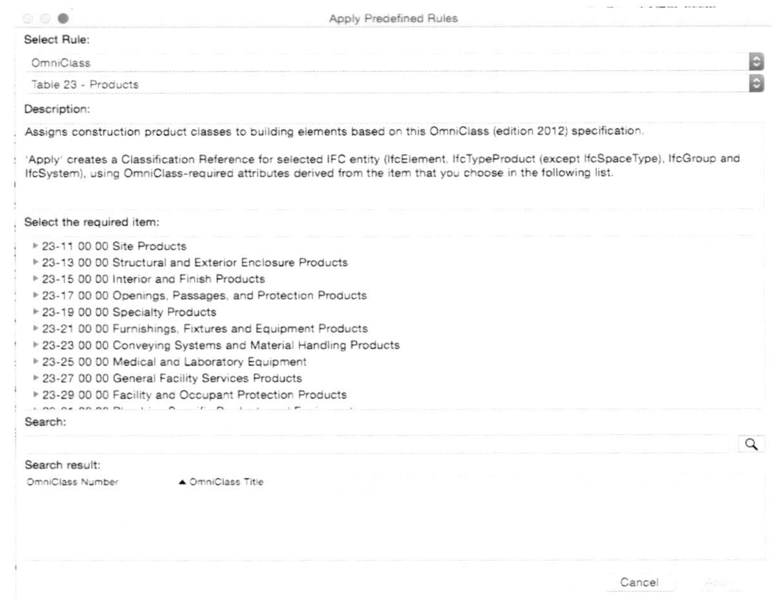

Figure 31 IFC Classification for Type

Requirement

Like Components, COBie data is only required for Types that are manufactured products with a preventative maintenance requirement. To summarize the requirements of NBIMS-US V3, Section A.4, the tables below identify the typical or expected list of included (Figure 32) and excluded (Figure 33) IfcTypeProducts in COBie

- Actuator Type
- Air Terminal Box Type
- Air Terminal Type
- Air To Air Heat Recovery Type
- Alarm Type
- Boiler Type
- Building Element Proxy Type
- Chiller Type
- Coil Type
- Compressor Type
- Condenser Type
- Controller Type
- Cooled Beam Type
- Cooling Tower Type
- Covering Type
- Damper Type
- Discrete Accessory Type
- Distribution Chamber Element Type
- Door Style
- Duct Silencer Type
- Electric Appliance Type
- Electric Flow Storage Device Type
- Electric Generator Type
- Electric Heater Type
- Electric Motor Type
- Electric Time Control Type
- Evaporative Cooler Type
- Evaporator Type
- Fan Type
- Filter Type
- Fire Suppression Terminal Type
- Flow Controller Type
- Flow Instrument Type
- Flow Meter Type
- Furnishing Element Type
- Gas Terminal Type
- Heat Exchanger Type
- Humidifier Type
- Lamp Type
- Light Fixture Type
- Mechanical Fastener Type
- Motor Connection Type
- Outlet Type
- Protective Device Type
- Pump Type
- Sanitary Terminal Type
- Sensor Type
- Space Heater Type
- Stack Terminal Type
- Switching Device Type
- System Furniture Element Type
- Tank Type
- Transformer Type
- Transport Element Type
- Tube Bundle Type
- Unitary Equipment Type
- Valve Type
- Waste Terminal Type
- Window Style

Figure 32 Typical IFC Type Products Included in COBie

Beam Type	Member Type
Cable Carrier Fitting Type	Pipe Fitting Type
Cable Carrier Segment Type	Pipe Segment Type
Cable Segment Type	Plate Type
Column Type	Railing Type
Curtain Wall Type	Ramp Flight Type
Duct Fitting Type	Slab Type
Duct Segment Type	Stair Flight Type
Element Type	Vibration Isolator Type
Fastener Type	Wall Type
Junction Box Type	

Figure 33 Typical IFC Type Products Excluded in COBie

Based on the requirements of NBIMS-US V3, identified in Figure 32 and Figure 33, the following IFC Type Products would only be required from an Architectural model for COBie:

- Building Element Proxy Type (i.e. Other Type)
- Covering Type (e.g. includes Ceilings, Cladding, Flooring and Roofing)
- Discrete Accessory Type (e.g. includes Shading Devices)
- Door Style
- Furnishing Element Type
- Lamp Type
- Light Fixture Type
- Roof Type
- Sanitary Terminal Type (e.g. includes Toilets, Baths, Sinks, Showers etc.)
- System Furniture Element Type
- Window Style

It should be noted that Covering Types where to have been excluded from the NBIMS-US V3 specification; however, they were included as a result of an oversight. It is likely that software systems using Covering objects will be required to push such information into Space.Attributes to be provided in a method consistent with the presentation of Room Finish Schedules.

Also, it should be noted that the East Dormitory project does not contain, and therefore was unable to test Discrete Accessory Types or System Furniture Element Types.

Requirements

The following Type fields are required as a minimum:

- Name
- Category
- Description
- AssetType

The following fields are not required by NBIMS-US V3, but are required by COBie-UK-2012:

- NominalLength
- NominalWidth
- NominalHeight

The following Type fields are not required to be provided during the design stage, since such information is completed by contractors:

- Manufacturer
- ModelNumber
- ModelReference
- ReplacementCost
- WarrantyGuarantorParts
- WarrantyDurationParts
- WarrantyGuarantorLabor
- WarrantyDurationLabor
- WarrantyDescription

The following Type fields are required if specified by the client. In practice, these fields have never been provided as they would be expected to be automatically populated based on manufacturer-compliant BIM templates:

- ExpectedLife
- Shape
- Size
- Color
- Finish
- Grade
- Material
- Constituents
- Features
- AccessibilityPerformance
- CodePerformance
- SustainabilityPerformance

All the above fields are completed in ARCHICAD's IFC Manager.

The following Type fields are automated by ARCHICAD:

- CreatedOn
- ExtSystem
- ExtObject
- ExtIdentifier

The following Type fields are automated by ARCHICAD at the point of export based on export unit settings (IFC Translation Setup):

- WarrantyDurationUnit
- DurationUnit

The following Type fields are produced by completing the required Contact and Space information respectively:

- CreatedBy
- Space Names

Issues

In order to complete the Type Name, Description and Category for the East Dormitory project, each Type had to be manually edited in the IFC Manager to achieve the desired output. AssetType also needed to manually edit items of Furniture where mapping could not predefine a value of Fixed or Movable.

Although the required manual editing was a straightforward process, the level of effort required is not realistic or manageable on a real life project where the design will most likely continue to evolve even after a COBie export has been created. It has been suggested that object naming conventions will suffice, this is largely irrelevant in ARCHICAD due to ARCHICAD's fully parametric approach to Component creation.

Edited Type data associated to a placed Component cannot be copied between files without losing the Type data.

Any item that is converted to Morph (see Component limitations / known issues) after any manual override of its Type data will revert to the automatic mapping and lose any manually inputted data including its Category (Classification Reference). The data would need to be recreated for these Types. Changing these elements also resulted in the labels for drawn output being lost and these needed to be recreated after conversion.

There is no currently no ability to preset information at Type level in ARCHICAD's Favorites. For example, many Components could be preset with AssetType and Category (Classification Reference) if this functionality was provided by ARCHICAD.

Another item that was a problem was at Space Types (IfcSpaceType) for export to COBie. Space Types are directly connected to Spaces (IfcSpace / Zone Stamp). It is difficult to see how this could be handled as this would affect import and export of this data.

It is also not currently possible to exclude Wall Types (IfcWallTypes) for export to COBie. Doors and Windows are embedded within Walls, so switching off Walls will result in the loss of these Types.

During testing the removal of excluded object types was accomplished by the COBie Plugin for bimServer.

NominalHeight, NominalLength and NominalWidth were set by default as IfcReal types of data in the version and build of ARCHICAD used for the East Dormitory project. This meant when mapping values, the units were shown in meters. In order to output the values as millimeters, it is necessary to create these 3 parameters as a different data type (for the East Dormitory project these were set as IfcLabels). Size is an IfcLabel so this was not affected. This issue was reported to GPAPHISOFT during this project and has been resolved in subsequent versions and builds.

Note that if commas are not required in NominalHeight, NominalLength, NominalWidth, Size or in any Description on an Apple iOS, then this must be altered in the System Preferences (Language & Region > Advanced… > Grouping) and not in ARCHICAD.

Best Practices

Name and Description had to be manually entered for each Type. By creating sophisticated mappings, this process could be somewhat streamlined. The mapping created ensured that each Type had been created and then the text was simply edited in the IFC Manager to the required values.

NominalLength, NominalWidth, NominalHeight and Size can be fully automated from the geometry in the model within ARCHICAD's IFC Scheme mapping. AssetType can also largely be automated for all Components with the exception of Furniture, which will need to be manually allocated as Fixed or Movable by the user.

Manufacturer, ModelNumber, WarrantyGuarantorParts, WarrantyGuarantorLabor, ReplacementCost, ExpectedLife, WarrantyDescription, ModelReference, Shape, Color, Finish, Grade, Material, Constituents, Features, AccessibilityPerformance, CodePerformance and SustainabilityPerformance can all have placeholder default values of 'n/a' set in ARCHICAD's IFC Scheme mapping. WarrantyDurationParts and WarrantyDurationLabor were also set with mapping to have default values of '0'.

An Interactive Schedule can be setup to ensure Type Names are unique. It is also helpful to have a schedule to compare the Type Name against the Component Description.

Impact

The logic of the relationship of Types to Components is an easy one to explain to Architects and ARCHICAD users. Types have more use than simply COBie and are key to robust models for quantification, for example. That said, it can be done as the East Dormitory project demonstrates but it takes time, effort, and extreme care to deliver it correctly.

Type management is one area that GRAPHISOFT need to give some further thought about how this process can be simplified for real life projects to make it realistic for all projects to deliver consistently.

COBie.SYSTEM

Overview

Systems provide the organizational structure for groups of COBie.Compnents. Systems are typically defined following early design by consulting engineers (or later by contractors and commissioning agents). The East Dormitory project did not add systems. Systems are largely to be considered for MEP Components although the UK approach, in BS 1192-4:2014, suggests all Components should be allocated to a System. COBie-UK-2012 has also suggested that there be a minimum number of systems in every COBie file.

Process

If Systems are required, then the process for creating them is similar to how Zone information is created from Spaces.

System data in ARCHICAD is created in the IFC Manager by dragging Components into the Assignments area into System folders created either in a template or manually by a user.

Requirements

The following System fields are required as a minimum:

- Name
- Category

The following System field is required if specified by the client:

- Description

The following Systems fields are automated by ARCHICAD:

- CreatedOn
- ExtSystem
- ExtObject
- ExtIdentifier

The following Space fields are produced by completing the required Contact and Component information, respectively:

- CreatedBy
- ComponentNames

Issues

The process of managing Systems gets tricky when the design changes after the System information has been created. For example, if one wants to move Furniture to another floor in ARCHICAD, it is a simple case of cut and paste. However, as soon as the user does this, the System information is lost and needs to be recreated.

COBie.ASSEMBLY and COBie.CONNECTION

Architects are not responsible for information in these two COBie Tabs. If information is required to be provided, it is not needed until engineering systems have been designed, or shop drawings or commissioning documents are produced.

COBie.Assembly records allow large equipment to be separated into constituent parts. COBie.Assembly should only be applied to large, complex components such as chillers or electrical distribution panels.

COBie.Connection identifies the logical control of one COBie.Component over another. For example, a switch or valve would be identified in COBie.Connection with its downstream terminals, fixtures, or controlled equipment.

COBie.SPARE, COBie.RESOURCE, and COBie.JOB

Architects are not responsible for information in these three COBie Tabs. Information in these three COBie Tabs is provided by the construction contractor, subcontractors, and commissioning agents prior to beneficial occupancy of the building.

COBie.Spare delivers information about on-site spare parts, replacement parts, consumables, and lubricants. COBie.Resource(s) are those material, labor, and equipment requirements needed to perform COBie.Job(s).

COBie.ISSUE and COBie.IMPACT

NBIMS-US V3 depreciates the use of these COBie Tabs entirely. The requirement for these Tabs was driven only by COBie-UK-2012. The ability to require to deliver such information, and the ability to validate the quality remain open questions.

COBIE.DOCUMENT

With the exception of Directory, File, and Reference, all this information is populated from other information already inputted into COBie. For architects, COBie.Document requirements are only required by specified by a client.

The design of COBie does, however, allow architects to use the COBie.Document data structure to help organize the entire set of building data. That is because an architect's list of "required" documents may be used as the basis for construction submittal and approval processes. Once approved, these submittals, become the attachments reference within COBie.Document.

The COBie.Document records provided in the East Dormitory COBie files were not produced by ARCHICAD, but by a feature of the COBie Plugin for bimServer. The COBie Plugin assumes that the contractor will be required to provide, at a minimum, once set of manufacturer's product data for each COBie.Type. Architects not wanting to include such information in files produced using the COBie Plugin may delete any rows so produced by the Plugin.

COBIE.ATTRIBUTE

The quality standard for COBie data complying with NBIMS-US is that the format of the COBie file (either SPFF or xlsx) be correct, and that the content of the COBie file exactly match the information provided on drawing schedules. By applying such a quality controls standard it is quite straightforward to understand the purpose of As-Designed COBie as the combined set of all the room finish and product schedules on the drawings. Furthermore, by specifying such a quality standard. the requirements of the contents of COBie can follow industry practice – rather than being forced into specific (limited) properties that do not fully reflect the products installed in the building.

The Attribute sheet is populated with the model properties that are not covered on the other sheets. COBie-UK requirements go beyond providing what is normally produced as part of drawing schedules. BS 1192-4:2014 identifies additional Attributes for Facility (Tables 8, 9 and 10), Space (Tables 11 and 12), Component (Tables 13, 14 and 16) and

Type (Table 15). These additional fields would appear on the Attributes sheet. The Attribute sheet contains any Attributes contained in the model other than those populated on other sheets.

The reader may recall that one of the ways that COBie requirements may be customized is for the client to specify the list of attributes needed for a given object. In the United States, such requirements might be identified in a "COBie Guide" or commentary that accompanies NBIMS-US V3.

In the UK, attributes are set out by clients in an Asset Information Requirements (AIR) document, which would then be read in conjunction with an Employer's Information Requirements (EIR) document. In addition to BS 1192-4:2014, standard Attributes are identified in ISO 16739:2013 (IFC2x3), so these would be the best place to create a list of Attributes for handover. It is likely that some further Attributes would also be required to provide a full picture of information required to manage the preventative maintenance of the asset.

Additional Attributes added at the Project (IfcProject) and Site (IfcSite) levels in ARCHICAD's IFC Manager are not available within the generated COBie output for the East Dormitory project. This is not an ARCHICAD issue, as the data is exported in the IFC file, but one for tools producing the COBie output outside of ARCHICAD. This would be an issue if the additional Attributes, identified by BS 1192-4:2014, were required for a COBie output. In particular, RefElevation, RefLongitude, RefLatitude and LandTitleNumber at Site level are not available. Additional Attributes added at the Project level are also not in the COBie output. These include BBATemplateAuthor, BBATemplateCreationDate and BBATemplateVersion.

COBIE.COORDINATE

The COBie.Coordinate Tab is a remnant from earlier, and largely depreciated COBie version. It allows the identification of points, lines, and boxes. As such this limited geometry was recoginized as not being relevant to the actual function of COBie. Since the COBie business case, as defined in NBIMS-US V3, only requires the concept of "spatial containment," i.e. "Pump-5 in Room-3," COBie.Coordinate cannot be justified. Furthermore, even COBie.Spaces in the simpliest of buildings cannot be defined by a single box. Any "L-shaped" spaces, or spaces with alcoves or closets breaks the COBie.Coordinate Tab specification. While

COBie.Coordinate data is provided by ARCHICAD, it should not be used in practice.

The information that populates the Coordinate sheet is automatically created at export so a user doesn't need to create any data manually. However, in order for this information to be correct the model itself must be in the correct project location within the environment from which the final output is created. This typically means that the model should be exported from a Site Model located in the correct location according the client's requirements.

AUTOMATING IFC DATA

Many users of BIM authoring tools are still focused on producing the same deliverables they have always produced but in a more efficient way. This means that even if they have adopted an approach to producing data much of it will be focused on native data fields within their chosen authoring tools. This is fine if the only output is a drawing or a schedule, but if others want to use published data for other purposes in a consistent manner on every project, irrelevant of who the model author is, then data needs to be built around a common open standard.

In BIM terms, the obvious data standard to align with is IFC (ISO 16739:2013) which is an open international standard developed by buildingSMART International. The current release being used in practice is IFC2x3 (IFC4 is now ready for implementation by the vendors so we are likely to see this implemented fully over the next couple of years). Some other reasons though to create an approach to standardizing data fields are as follows:

- To automate data from geometry wherever possible, reducing opportunities for errors with manual input
- To collaborate with other model authors without the need to map data between platforms or even authors using the same tool
- To provide consistency for all tools and objects, across all project files and ultimately across all projects
- To allow model checking to be built around a single consistent standard
- To deliver COBie and the requirements of BS 1192-4:2014 efficiently and effectively (required for centrally funded UK government projects from April 1st 2016 for "Level 2 BIM" projects)
- To allow consistent scheduling of data fields across a variety of modelling tools
- To not need to rely on all imported geometry to have consistent data fields
- To create a single approach to where data is created making training simpler for staff
- To develop a business around "OPEN BIM" that is more than just sharing geometry and/or unstructured data
- To divorce the silo mentality around authoring tools

- To future proof a business approach to data around an open standard

In ARCHICAD we have the ability to focus on IFC data by using ARCHICAD's IFC Scheme Mapping. The second author of this book has written before on this subject in his blog entitled "ARCHICAD 18: Join the IFC and COBie flow – Part 2 – Data Mapping "Through this mapping process designers may better understand how people use this and work out ways to further automate their collaborative design workflows. Doing so has the side benefit of delivering well-structured open-data with no additional work.

The simple scenario described in the following paragraphs demonstrates the need for mapping. ARCHICAD 19's out-of-the-box "Bar Stool 19" object is shown in Figure 34.

The same size object (400 long x 400 wide x 815 high) has been placed multiple times but each with a different Style associated. There is one instance of Style 1, two instances of Style 2, three instances of Style 3, four instances of Style 4, and five instances of Style 5. There are 15 instances (or Components/Elements) in total. When thinking about COBie each "Style" is a COBie.Type and the instances are COBie.Component(s).

Figure 34 ARCHICAD Bar Stool Placement

Figure 35 is an example of the information available for thee out-of-the-box objects without mapping.

Figure 35 IFC Data Manager Barstool without Mapping

The Name in this case is unique as each one was placed. However, if the designer duplicates any of the objects the IDs lose their uniqueness (a weakness of ARCHICAD). No Description for the object is automated with the out-of-the-box template and there are no placeholders or other linked values as illustrated in Figure 36.

The objects all appear as a single Type with the Type Name using only the Library Part Name. No other values for the object are automated with the out-of-the-box template and there are no placeholders or other linked values. On the whole, the out-of-the-box template solution cannot take advantage of the information we know about the project. The impact is, for example, that a cost consultant or contractor would assume that there are multiple instances of exactly the same bar stool. The cost for different types of objects, however, may be different so the estimator would need to re-create uniqueness for each different Type.

So without mapping the out-of-the-box scenario is going to give us a number of issues. The most important reason is that without some type of mapping project team members cannot easily use the information in your model to produce a reliable quantification of the model's objects. A side effect of having a higher quality model that includes more than geometry is that COBie data may be exported with little additional effort.

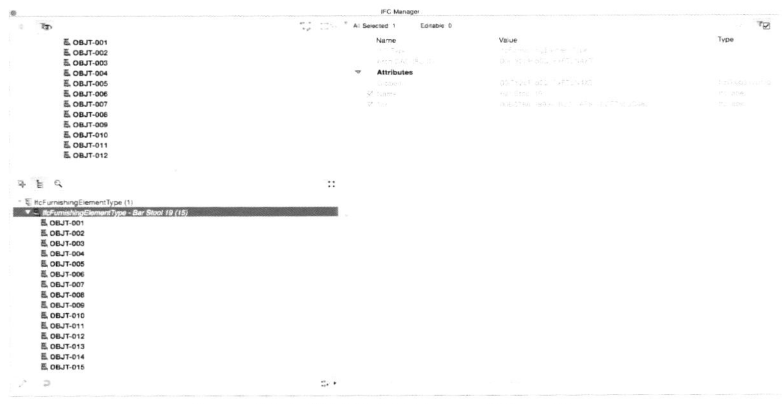

Figure 36 ARCHICAD Barstool Type Data

Taking the same objects but applying the IFC mapping we have developed we get the results show in Figure 37.

Figure 37 ARCHICAD Barstool Component Data

The Component Description is fully automated from the geometry selected by the user. Placeholders are also created for COBie Component data (AssetIdentifier, InstallationDate, TagNumber, WarrantyStartDate, BarCode and SerialNumber). Only the ID (mapped to the Component Name), and in some cases the Classification References (a large amount can be pre-setup in Favorites), needs to be managed by the user.

Figure 38 IFC Manager Barstool Type Data

The crucial difference here compared to the out-of-the-box scenario is that the 5 different Types each have a unique Type Name. This means that these can be used for the COBie Type sheet but also we can use these Types to develop an approach to quantification.

The standard data structure for COBie is derived from the underlying COBie MVD and the mapping between STEP File format and spreadsheet format found in NBIMS-US V3. As we have seen earlier in this book, that standard design information does not necessarily guarantee that the information in the file will be useful to those receiving the information. The criteria used when evaluating the data is that the data has to match the drawings. To accomplish this, Type.Name(s) should be fully human-readable and exactly match the information on equivalent drawing schedules. For example, COBie.Type.Name should be of the form Bar Stool Type 1, Bar Stool Type 2, Bar Stool Type 3 etc. Currently this can be achieved manually, however it simply isn't currently practical for larger jobs since the ARCHICAD workflow for Type data cannot be preconfigured in ARCHICAD.

If one were to develop a consistent naming schema for COBie.Component, as identified in the US COBie Guide [East 2012a], it would be based on the following fields: Type Name, Component Space Name, and Component Space Count.

Once each Type.Name is unique, the names of the individual components should be created using the following formula: Component.Name = Type.Name +Component.Space. If for example, there is a specific type of sink called "LAV-A", then an example component of that sink in a washroom whose COBie.Space.Name is

"205" would be "LAV-A-205". When there are more than one instance of a given type of component in a space an additional "counter" is needed to ensure each individual COBie.Component.Name is unique. With this final criteria, the name of the third sink of type "LAV-A" to be placed within room "205" would be "LAV-A-205-3". At this time, it is not possible to automate this process or identified equipment that is within walls or above ceilings. More information about the best practices regarding naming within ARCHICAD are available from Bond Bryan Architects [Jackson 2015e].

Developing a mapping approach takes considerable thought and time. Also the process should be an iterative one that begins with a small example and is tested and checked. Once the process has been tested and applied across a practice, there is a real opportunity for standardization, automation, and the production of high quality valuable data.

CLASSIFICATION REFERENCES

Benefits

The following are some of the benefits of using a Classification Reference:

- To allow downstream users to find information without manually finding and counting building elements
- To support different classification system for different users (provided an XML file is available)
- To reduce individual data entry time and improve data quality
- To ensure consistent application of information across multiple users and projects
- To apply appropriate classification to different base information (e.g. to Project, Site, Building, Floor, Space, Zone, Component, Type, System Group, Actors, Space Occupants and Time Series Schedules)

Figure 39 shows where the results of classification appear on a Space Schedule. COBie (and the underlying IFC) require two parts to a given classification. The first part is the numbered reference, the second part is the name for the classification.

Templates

The ARCHICAD 18 and 19 UKI templates currently contain the following classifications:

- CAWS
- COBie Floor and Zone Categories
- Concept Design BIM 2010 (US GSA)
- OmniClass
- RICS NRM1
- Uniclass 1.4
- Uniclass 2
- Uniclass 2015 (from Update 6000)

The Concept Design BIM 2010 (US GSA) and OmniClass are predominantly for the US market with the others available for use in the

UK (and elsewhere if required). MasterFormat and UniFormat are other classification systems available from GRAPHISOFT.

COBie Spaces

Name	Description	Classification ItemReference	Classification Name	Floor Name	Usable Height	Gross Area	Net Area
101	EXERCISE ROOM	13-33 17 11 11	Exercise Room	LEVEL ONE	2750	60.7	60.7
102	TOILET	13-65 13 13	Toilet Space	LEVEL ONE	2440	4.2	4.2
103	TOILET	13-65 13 13	Toilet Space	LEVEL ONE	2440	4.2	4.2
104	STORAGE	13-63 13 13	Closet	LEVEL ONE	2750	10.4	10.4
105	MECHANICAL ROOM	13-23 19 00	Utility Equipment Room	LEVEL ONE	2750	38.6	38.6
106	BATH	13-65 13 00	Bathroom	LEVEL ONE	2440	4.0	4.0
107	CLOSET	13-63 13 13	Closet	LEVEL ONE	2440	2.9	2.9
108	BREAK ROOM	13-57 17 13	Break Room	LEVEL ONE	2440	15.8	15.8
109	ELEC/TEL	13-23 19 27	Electrical Room	LEVEL ONE	2750	4.0	4.0
110	LAUNDRY	13-65 17 00	Laundry Room	LEVEL ONE	2440	12.5	12.5
111	STORAGE	13-63 13 13	Closet	LEVEL ONE	2750	5.2	5.2
112	KITCHEN	13-57 13 13	Kitchen Space	LEVEL ONE	2750	37.8	37.8
113	FOOD STORAGE	13-63 13 13	Closet	LEVEL ONE	2750	13.6	13.6

Click an Element or Draw a Selection Area. Press and Hold Ctrl+Shift to Toggle Element/Sub-Element Selection

Figure 39 Space Schedule Showing Classification

Structure

A Classification Reference is not a single piece of data. It contains seven individual data elements. The first three data elements are listed at the top of the Classification Reference dialog in Figure 40, as follows:

1. Location: Location, where the external source (classification, document or library). This can be either human readable or computer interpretable. For electronic location normally given as an URL location string, however other ways of accessing external references may be established in an application scenario.
2. ItemReference: Identifier for the referenced item in the external source (classification, document or library). The internal reference can provide a computer interpretable pointer into electronic source.
3. Name: Optional name to further specify the reference. It can provide a human readable identifier (which does not necessarily need to have a counterpart in the internal structure of the document).

The remaining four Classification Reference data elements are listed at the top of the Classification Reference dialog in Figure 40, as follows:

4. Source: Source (or publisher) for this classification.

5. Edition: The edition or version of the classification system from which the classification notation is derived.
6. EditionDate: The date on which the edition of the classification used became valid.
7. Name: The name or label by which the classification used is normally known.

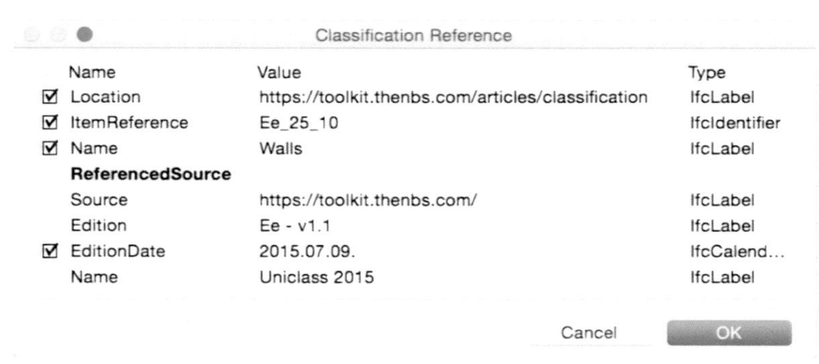

Figure 40 Classification Reference Dialog

EXPORTING COBIE FROM ARCHICAD

The exporting of an IFC2x3 model for COBie is very straightforward. GRAPHISOFT has created a standard translator for COBie. This is named "COBie 2 Export" and complies with the buildingSMART Basic FM Handover Model View Definition (MVD).

As the Model View Definition is no longer the basis of the NBIM-U V3, it would be expected that ARCHICAD change their documentation to reflect the current basis of the COBie 2.4 specification that they have implemented ARCHICAD v19. The recommended amendment is that the translator is renamed 'COBie 2.4 MVD' for future versions.

The IFC Model Unit Settings may need to be altered depending on region and/or client requirements. The following are controlled in the IFC Model Unit Settings: Length Unit, Angle Unit, Area Unit, Volume Unit, Currency Unit and Time Unit.

The Properties exported from a model can also be controlled at export. The export settings allow "All properties" or "Current Scheme properties only."

CONCLUSION

This book illustrates the process of extracting building information from GRAPHISOFT ARCHICAD BIM models. The information delivered in this process did not require external concepts of design quality or abstract concepts of collaboration. The information delivered simply reflected the professional standard of care applied in a typical design process. The new standard of care applied to the software used during the design process can be simply stated as "data that matches drawings."

In this context, the production of COBie data aligns well with existing design processes. The delivery of COBie, therefore, should not be thought as something different from standard architectural practice.

This book demonstrated a two-step process of testing COBie deliverables. The first step was the automated, open-source verification of COBie-formatted building information based on NBIMS-US V3 requirements. The second step was the manual validation of COBie content against information provided on simulated contract drawings.

Based on this testing procedure, the authors have concluded that the producing a compliant architectural COBie File from GRAPHISOFT ARCHICAD is entirely possible.

With currently available tools, designers are required to understand the mappings between their design data and COBie requirements. As tools improve to the point that the information found on drawings is automatically checked and delivered, the effort of designers to export scheduled building components should decrease. This process is iterative requiring the ongoing contribution of standards developers, practitioners, and software companies. This book is the next step in that process.

The hope is the East Dormitory project and the findings presented in this book will help to inform the continued improvement of ARCHICAD and other design software over the coming years.

RECOMMENDATION

There are two primary recommendations arising from the development of the East Dormitory as a test project. One of these recommendations pertains to the design software used. The other recommendation is made for architects beginning to work with COBie data exports.

With regard to software developers, the lesson from this report is clear. The conference-room testing of design software for the purpose of COBie testing is only able to address the verification stage of testing. To ensure that users are able to effectively use the tools provided, hands-on, end-user software testing is required. The validation stage of testing ensures that real designers, creating real designs, can produce data that matches their drawings.

With regard to architects starting to deliver COBie data, there is also a clear lesson from the information provided in this book. While the acronym BIM typically refers to the geometric modeling of buildings, when thinking about COBie, the acronym should really be "Better Information Management." To effectively create COBie data, at least one person in each design office should be tasked with deeply learning the COBie standard, then creating standard office templates, classification references, and object libraries with default information already prepopulated. These individual should also be in the position to help kick-start projects from the start.

APPENDIX

During the development of this book, the second author compiled a series of notes regarding potential product enhancements to ARCHICAD 19 (Build 4013). These notes are listed in the paragraph below.

Contact

Contact information currently cannot be used in Interactive Schedules or Project Indexes within ARCHICAD. This would be useful in order to reuse data for other purposes such as Transmittal Sheets or Task Information Delivery Plans (TIDP).

It also would be useful to be able to schedule this data in order to replicate the COBie schedule for checking purposes.

Facility

Values for COBie.Facility cannot currently be created in Project Info and then mapped into IFC fields for Project, Site, or Building levels other than those that are automatically mapped by GRAPHISOFT. If that were done, then such fields, including Classification References (Category), would also be accessible from Project Info.

If this amendment were done, the user would be able to use Project Info to complete ALL Facility information required by COBie. Although the existing process works as is, this change is primarily required to simplify the process for users.

Project, Site, and Building information currently cannot be used in Interactive Schedules or Project Indexes within ARCHICAD. Having this information would allow the user to replicate the COBie schedule for checking purposes. In addition, the information could be used for other purposes at the design office

Floor

An item identified during the production of the Dormitory project was that ARCHICAD 'Height to Next' (Height) currently does not export from ARCHICAD version 19. It does, however, export from versions 17 & 18. GRAPHISOFT have confirmed this as a bug and it will be fixed in an upcoming update. Users should check the release notes provided with builds since build 4013 to see if the issue has been resolved.

For simplicity of use, it would be easier if the Description and Classification (Category) could be applied in the Story Settings dialogue.

While COBie only requires the Category (IfcClassificationReference) to appear in the Name field, we have found some software that will not recognize this value without an additional value in the Item Reference field. In the example above, notice we have added a matching value into the Item Reference field i.e. Name = Floor and Item Reference = Floor.

Care must be taken by users when renaming Floors in ARCHICAD's Story Settings. Users may rename Floors, unaware that they must also consider the Description and Category in the IFC Manager. Manual checking processes need to be put in place to ensure data is correct.

Description and Category (and other data from the IFC Manager at Floor level) currently cannot be used in Interactive Schedules or Project Indexes within ARCHICAD. This would be useful in order to reuse data for other purposes.

It also would be useful to be able to schedule this data in order to replicate the COBie schedule for checking purposes.

Space

The naming of the Zone Stamp is unfortunate in terms of explanation to users. A number of requests to GRAPHISOFT have been made to rename this tool in future versions of ARCHICAD. This will make the alignment between ARCHICAD and open workflows more consistent and easier to teach to users.

Creating Spaces which have multiple heights is both limited by ARCHICAD and IFC/COBie's ability to only accept a single Space Height value. In some cases, it is necessary to split Spaces where significant Height differences occur - for example, Atrium spaces connected to lower height Spaces.

In COBie, there is a critical relationship between Spaces and the products and equipment (i.e. COBie.Component). Products and equipment that are installed within walls must be assigned to the space from which they are used.

For the East Dormitory project there were a number of Components embedded within Walls (this included the Soap Dishes and Medicine Cabinets). In order for these Components to be associated to the relevant Spaces it was necessary to adjust the Spaces to incorporate these Components. This adjusted the floor area of each of the Spaces, in this case by 0.1m^2, but this was a workaround to enable the rules of COBie to be passed. This workaround would need to be flagged up to a client particularly if the client was utilizing this data for other purposes, such as calculating

Zone

The process of allocating Spaces to Zones is fairly straightforward within ARCHICAD. However, the workflow becomes an issue as the design changes after the allocation process has taken place. For example, moving a Space (i.e. cutting and pasting a Zone Stamp for example) from one Floor to another will result in the loss of the relationship between Space and Zone.

The IFC Manager also lacks the ability to see whether a Space has been allocated to a Zone. The process requires careful management by the user and it is easy to fail to allocate a Space to a Zone.

There is a need to have a feature to see which Spaces are unallocated to Zones (IFC Zones) within the IFC Manager. Find & Select can be utilized to search for unassigned Space to Zones.

There is no way to preconfigure the relationship of a Space (Zone Stamp) to a Zone. Ideally Space and Zone relationships could be set up as part of configuring ARCHICAD Favorites within an office ARCHICAD template.

Zone data cannot be scheduled in Interactive Schedules or Project Indexes in ARCHICAD other than the Name field. This would be useful in order to reuse data for other purposes. It also would be useful to be able to schedule this data in order to replicate the COBie schedule for checking purposes.

Type

Type data cannot be scheduled in Interactive Schedules or Project Indexes in ARCHICAD other than the Name field. This would be useful in order to reuse data for other purposes. It also would be useful to be able to schedule this data in order to replicate the COBie schedule for checking purposes.

As Type data cannot currently be scheduled, it results in more Attribute data being managed at Component level by the user in ARCHICAD. This can lead to a greater chance of error with managing data for projects.

Classification References

The following are limitations with using IFC Classification References (of course there may be more):

1. Only one Classification Reference from one table can be assigned to an Element/Component. For example, a user can not choose 2 things from the Elements table of any of the Uniclass or OmniClass tables. Note: COBie would only accept one classification per Element so there is also a potential limitation of COBie not just IFC Classification References
2. Classification References for Component (Element/Object) and Space can only currently be scheduled in ARCHICAD.
3. No other Classification References can be scheduled (e.g. Facility, Floor, Type, System or Zone).
4. Classification References cannot be used with the out-of-the-box ARCHICAD labels except at Element/Component level.
5. Classification References cannot be applied to Building Materials or individual skin components of a Composite or Profile construction (this is to do with restrictions of IFC2x3 and likely to be improved with IFC4 and aligned future ARCHICAD improvements).

REFERENCES

Bogen, Chris. (2015). "COBie Plugin for bimServer.org", Prairie Sky Consulting, GitHub Repository. (https://github.com/opensource BIM/COBie-plugins/releases).

Bogen, Chris and East, E. William. (2015). "COBie Plugin for bimServer.org: A Quick Start Guide," Prairie Sky Consulting, ISBN 9781329111745. (http://www.lulu.com/shop/chris-bogen-and-e-william-east/cobie-plugin-for-bimserver-a-quick-start-guide/ebook/product-22300824.html).

Bogen, Chris. (2016). "COBie QCReporter Command Line Tool," Prairie Sky Consulting, GitHub Repository. (https://github.com/OhmSweetOhm/CobieQcReporter/releases).

buildingSMART international. (2013). "Industry Foundation Classes, Release 4 (IFC4)", Section 5.4.3.45. (http://www.buildingsmart-tech.org/ifc/IFC4/final/html/link/ifcspace.htm).

Chipman, Tim. (2013). "Model View Definition, Industry Foundation Classes – Construction Operations Building Information Exchange," buildingSMART alliance, National Institute of Building Sciences. (http://docs.buildingsmartalliance.org/MVD_COBIE/).

Creative Commons. (2016). "Attribution-ShareAlike 4.0 International." (https://creativecommons.org/licenses/by-sa/4.0/).

East, E. William. (2016). "Software Testing," Prairie Sky Consulting. (http://www.prairieskyconsulting.com/testing.htm).

East, E. William. (2014). "Construction Operations Building Information Exchange: Means and Methods," buildingSMART alliance, National Institute of Building Sciences. (https://www.nibs.org/?page=bsa_cobiemm).

East, E. William. (2013). "Barracks 101," buildingSMART alliance, National Institute of Building Sciences. (http://projects.buildingsmartalliance.org/files/?artifact_id=6701).

East, E. William. (2012). "COBie Guide," U.S. Army Corps of Engineers, Engineering Research and Development Center, Public Release 2. (http://projects.buildingsmartalliance.org/files/?artifact_id=4994).

East, E. William. (2012). "Construction Operations Building information exchange," DRAFT Unified Facilities Guide Specification, UFGS 01 79 00. (http://projects.buildingsmartalliance.org/files/?artifact_id=6674).

East, E. William. (2009). "Reducing Handover Documentation Costs using COBie," A presentation to the Associate of General Contractors, U.S. Army, Corps of Engineers, Engineer Research and Development Center.

East, E. William. (2007). "Construction Operations Building Information Exchange (COBIE) Requirements Definition and Pilot Implementation Standard," U.S. Army, Corps of Engineers, Engineer Research and Development Center, TR 07/30. (http://www.dtic.mil/cgi-bin/GetTRDoc?Location=U2&doc=GetTRDoc.pdf&AD=ADA491932).

Jackson, Robert. (2015). "Understanding Element Classification for IFC 2×3 exchange in ARCHICAD 18/19 – Part 1 – Relationships." (http://bimblog.bondbryan.com/understanding-element-classification-for-ifc2x3-exchange-in-archicad-1819-part-1-relationships/#more-3162).

Jackson, Robert. (2015). "Understanding Element Classification for IFC 2×3 exchange in ARCHICAD 19 – Part 2 – ARCHICAD Type and Library Parts." (http://bimblog.bondbryan.com/understanding-element-classification-for-ifc2x3-exchange-in-archicad-19-part-2-archicad-type-and-library-parts/#more-3276).

Jackson, Robert. (2015). "Understanding Element Classification for IFC 2×3 exchange in ARCHICAD 18/19 – Part 3 – Predefined Types." (http://bimblog.bondbryan.com/understanding-element-classification-for-ifc2x3-exchange-in-archicad-1819-part-3-predefined-types/#more-3315).

Jackson, Robert. (2015). "IFC 2×3 element and type classification – The plain language A-Z list". (http://bimblog.bondbryan.com/ifc-2x3-element-and-type-classification-the-plain-language-a-z-list/#more-3667)

Jackson, Robert. (2015). "Developing a standardized and automated approach to IFC data fields." (http://bimblog.bondbryan.com/developing-a-standardised-and-automated-approach-to-ifc-data-fields/).

GRAPHISOFT. (2015). "GRAPHISOFT ARCHICAD and COBie2: How to Prepare your ARCHICAD 19 Project for COBie2 Documentation." (http://www.graphisoft.com/ftp/techsupport/documentation/IFC/ARCHICAD19%20and%20COBie2.pdf).

National Institute of Building Sciences. (2015). "United States National Building Information Model Standard, Version 3." (https://www.nationalbimstandard.org/).
National Institute of Building Sciences. (2012). "Construction Operations Building information exchange (COBie), buildingSMART Alliance." (https://www.nibs.org/?page=bsa_cobie).
University of Florida, Gainesville. (2016). "The COBie Academy." (http://pd.dce.ufl.edu/programs--courses/cobie-academy/).

To gain access to guides for ARCHICAD from version 16 onwards available, visit: http://www.graphisoft.com/downloads/addons/ifc/index.html

To learn more about COBie, please visit The COBie Academy @ http://pd.dce.ufl.edu/programs--courses/cobie-academy/

All links checked for validity as of July 2016.